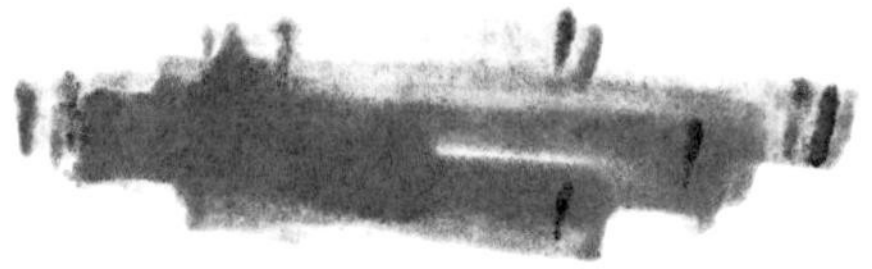

TELE-SATELLITE

Editorial: Steve Parker, Ian Graham
Design: David West
Children's Book Design
Illustrator: Alex Pang
Picture research: Cecilia Weston-Baker
Consultant: Tim Furniss, *Flight* magazine

Created and designed by
N.W. Books
70 Old Compton Street
London W1

First published in the United States by
Gloucester Press
387 Park Avenue South
New York, NY 10016

Printed in Belgium

Library of Congress Cataloging-in-Publication Data

Irvine, Matt.
Telesatellite/by Matt Irvine.
p. cm. -- (Engineers at work)
Includes index.
Summary: Discusses the role telecommunications satellites have played in the communications revolution, how they transmit telephone calls, TV shows, and weather information across the globe, how they stay in orbit, receive repairs, and are used by the military.
ISBN 0-531-17157-4
1. Artificial satellites in telecommunication--juveniles literature. [1. Artificial satellites in telecommunication.]
I. Title. II. Series.
TK5014.178 1989
621.382'5--dc20 89-31560
CIP
AC

Contents

ENGINEERS AT WORK

TELE-SATELLITE

MAT IRVINE

GLOUCESTER PRESS
New York · London · Toronto · Sydney

4 COMSAT!

The idea of placing an object in orbit around the Earth has occupied the minds of scientists for many years. Space projects have been discussed seriously since at least the beginning of this century, but it was more than half a century later before the "Space Age" became a reality.

The world's first artificial satellite was launched by the Soviet Union on October 4, 1957. It was called *Sputnik 1*. It was a simple metal sphere 58 cm (23 inches) in diameter, containing a radio transmitter. Sputnik 1 survived in orbit for only three months before it reentered the Earth's atmosphere and burned up. But this simple craft began a worldwide industry that now involves virtually every country.

Satellites have become much more complex since then. Some satellites are used for scientific projects. Others monitor weather patterns and examine the oceans or the face of the Earth itself. Today, many people are familiar with the "telesats" that are used to relay television and radio programs and telephone calls around the world. They are called communications satellites or "comsats."

Most domestic broadcasting still uses tall radio masts. A satellite can be thought of as an extremely tall radio mast which therefore has a far greater range.

Some early comsats were large metallized balloons designed to reflect radio waves. As they merely reflected signals, they were described as "passive."

Solar array

Comsats require electrical power for their receivers and transmitters. All comsats generate this power from solar cells. These convert sunlight into electricity.

A modern communications satellite is an extremely complex piece of equipment. Early comsats such as *Intelsat 1,* better known as "Early Bird," could carry up to 240 simultaneous telephone calls. Today's comsats such as *Olympus 1* (shown here) can each carry over 30,000 transmissions.

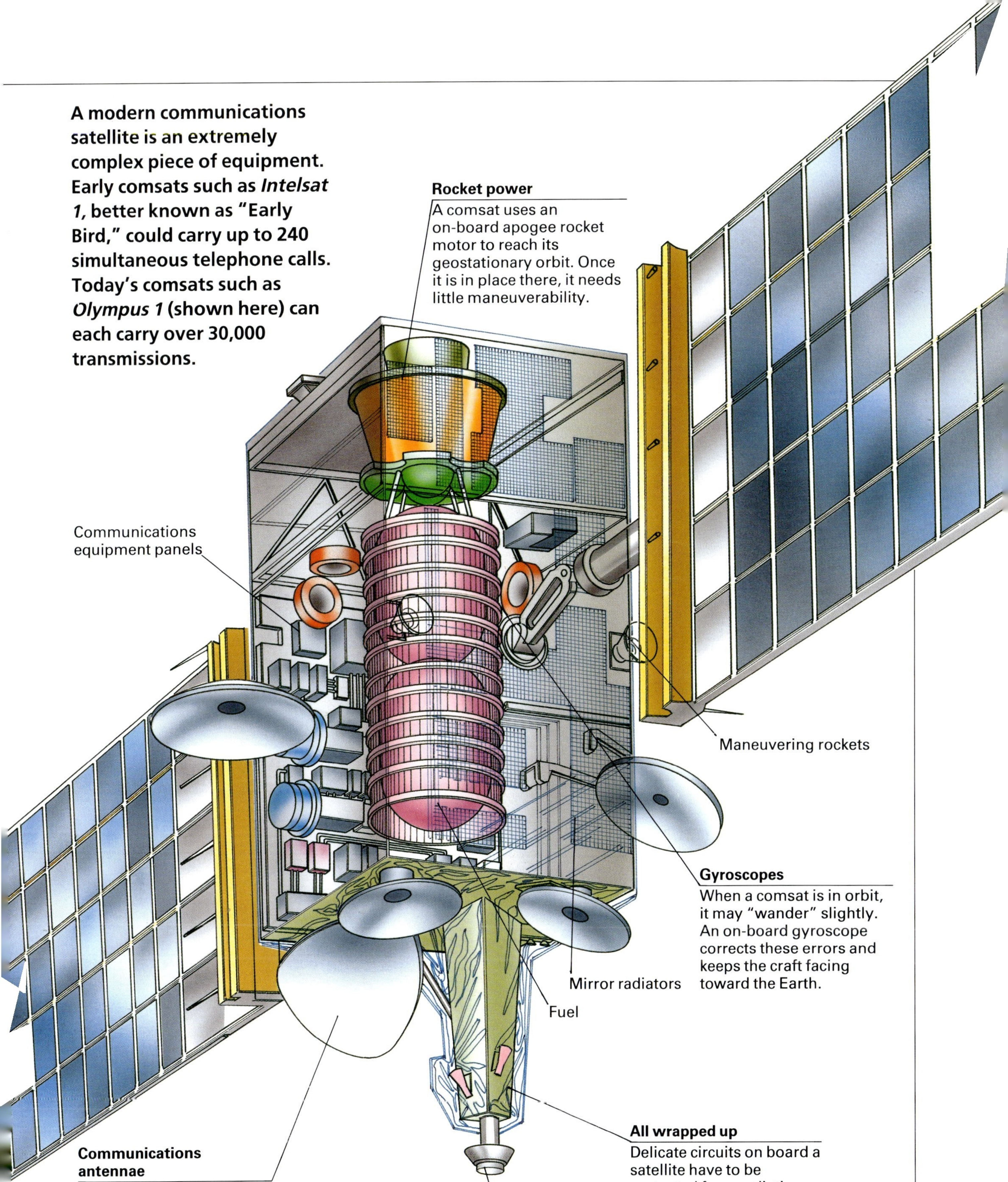

Rocket power
A comsat uses an on-board apogee rocket motor to reach its geostationary orbit. Once it is in place there, it needs little maneuverability.

Gyroscopes
When a comsat is in orbit, it may "wander" slightly. An on-board gyroscope corrects these errors and keeps the craft facing toward the Earth.

All wrapped up
Delicate circuits on board a satellite have to be protected from radiation from the Sun by thermal blankets of metal foil.

Communications antennae
Comsats must receive and re-transmit many different types of signals. This is why they do not have a single receiving dish or antenna. Instead, several dishes are angled in different directions.

6 ELECTRONIC OFFICE

All companies, whether they are large or small, rely on telecommunications. A small company may only have a single telephone line. Larger companies and organizations may have what amounts to a small telephone exchange. Through these lines they can keep in touch with suppliers, customers or subsidiary companies at the touch of a few buttons. The electronic signals travel almost at the speed of light.

Information can be routed into an office and out again in a wide range of forms, including voice communications, computer data, telex messages, and television pictures. Telephone lines, linked to worldwide communications networks by cables and satellites, now carry most of this information.

Telephone
The most popular form of electronic communication. Electronics and styles have changed, but they still work in the same way as the first telephones.

Telex
A system for sending typewritten messages by telephone or by radio. Messages are sent from a typewriter keyboard and received by teleprinter.

Computer
Computers are now common in both office and home. They can be linked to other computers via telephone lines by using a device called a modem.

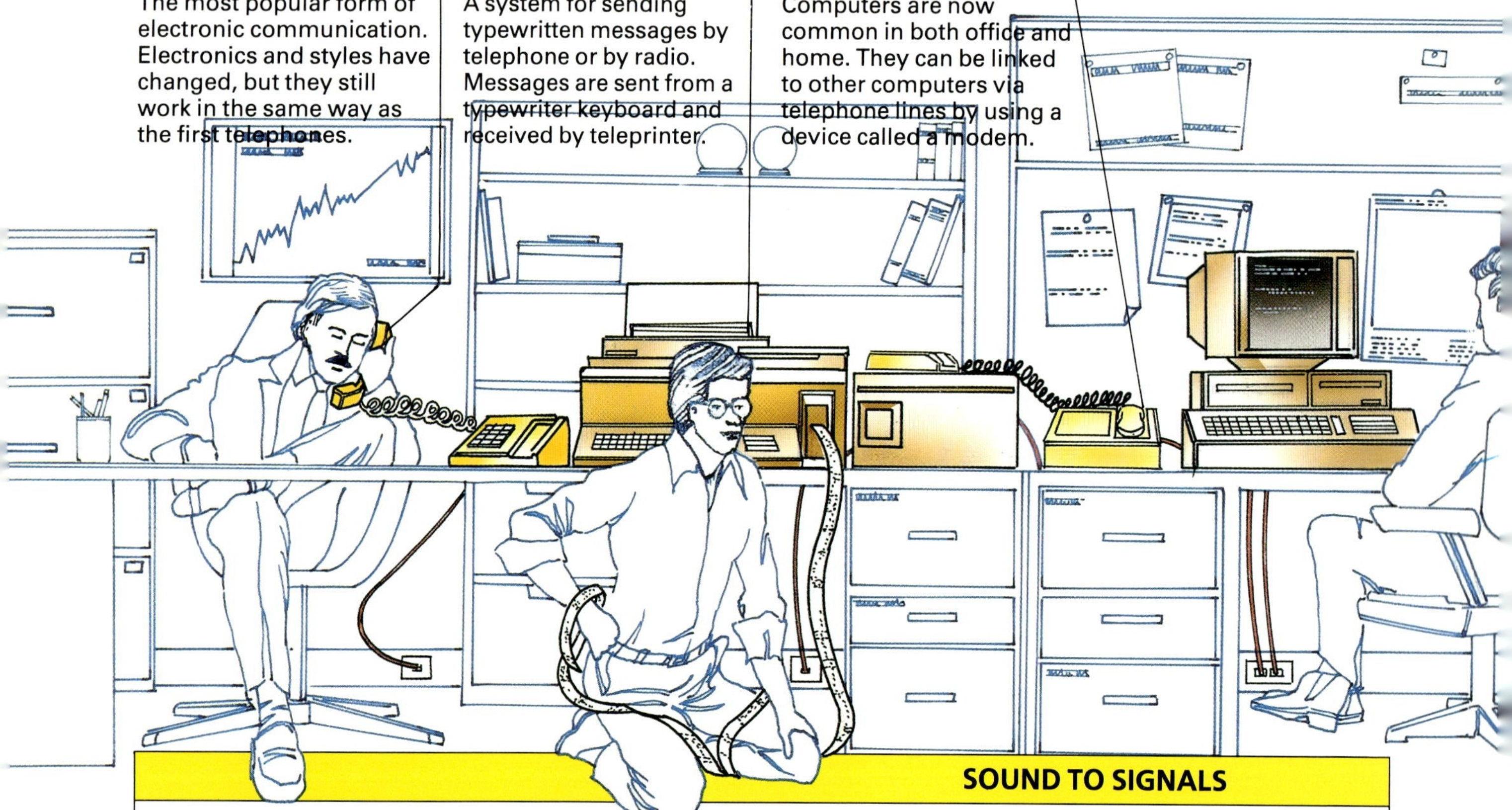

SOUND TO SIGNALS

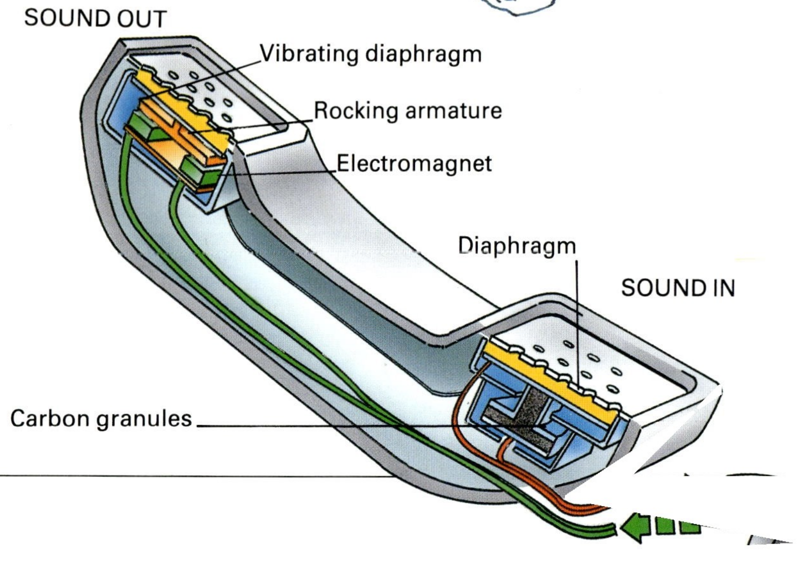

The telephone mouthpiece is a small microphone. It converts your voice into electrical signals. When you speak, the sound waves vibrate a thin sheet of metal called a diaphragm. This compresses tiny carbon granules underneath it and generates a small electric current which varies with the sounds.

All the electronic equipment that is now available means that we are rapidly approaching a situation where the office and the home can be the same place. Many people already work from home using the telephone lines to connect them with the outside world. Besides the telephone, computers and fax machines connected to ordinary telephone lines send and receive messages.

Most telephone lines use metal wires, but these are being replaced by fiber-optic cables, which carry messages as pulses of light. They are almost free from interference over long distances. A single optical cable can carry more messages than a wire ten times as thick.

Fax

The Facsimile Machine or "Fax" is the latest telecommunications aid for the office. Words, drawings, diagrams, in fact anything drawn or printed on paper, can be sent by a fax machine linked to a telephone.

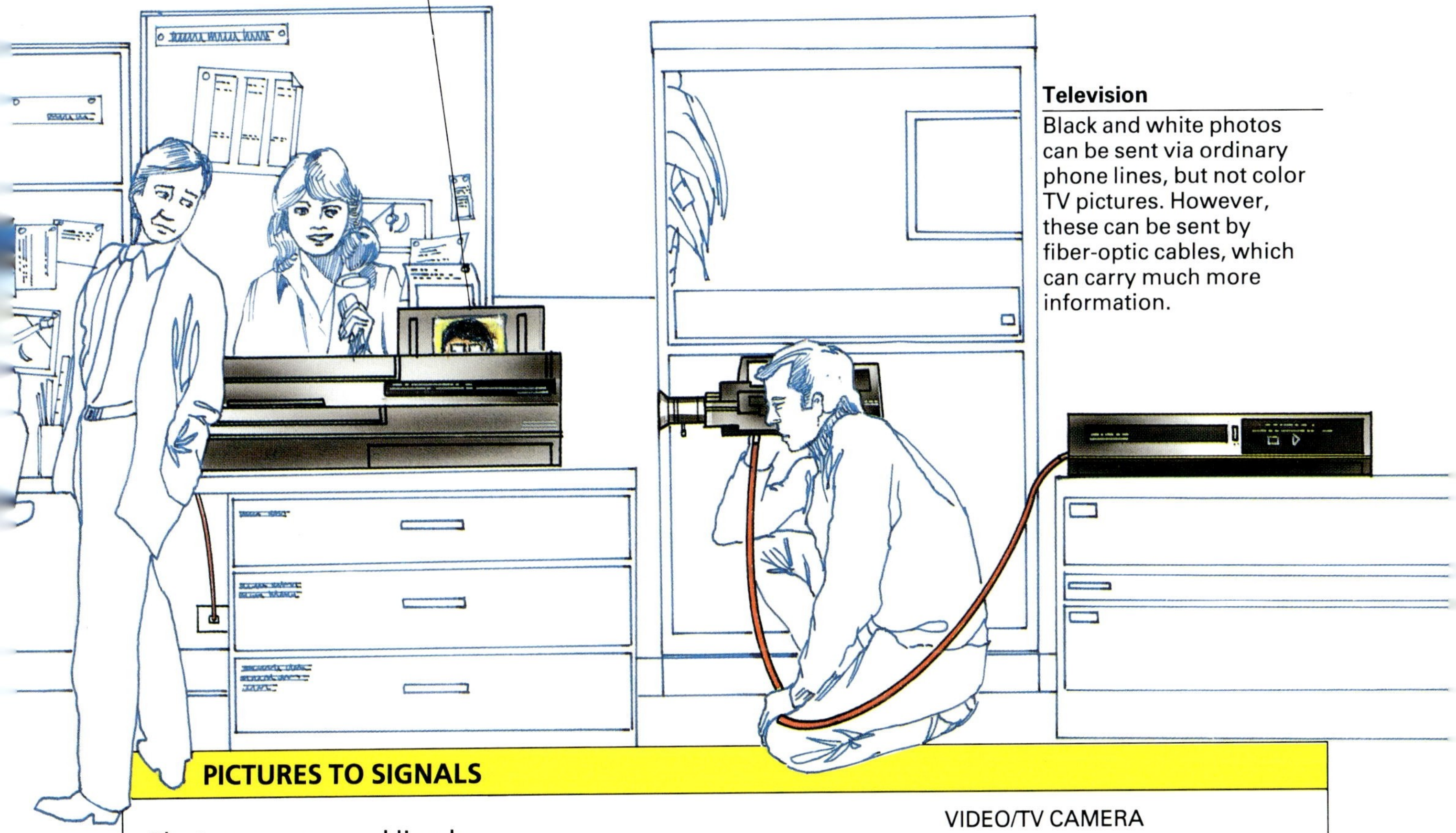

Television

Black and white photos can be sent via ordinary phone lines, but not color TV pictures. However, these can be sent by fiber-optic cables, which can carry much more information.

PICTURES TO SIGNALS

Photos are scanned line by line. The light and dark areas are converted into electrical signals. A similar machine receives and prints the photo in the same way. Television cameras scan at much higher speeds and use microchips called CCDs (Charge Coupled Devices) to change pictures into electrical signals.

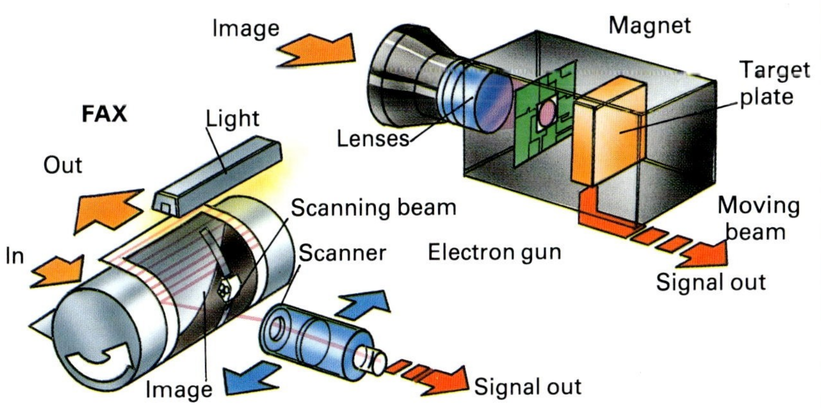

SENDING A MESSAGE

Electronic signals can travel around the world in a variety of ways. Radio broadcasts travel in three different forms according to their wavelength. Long waves follow the curvature of the Earth. Medium and short waves bounce off a layer of the upper atmosphere, called the ionosphere. The quality of the signals received changes depending on the condition of this layer. Usually, the signals travel farther at night. VHF radio signals have a very short wavelength and they only travel in a straight line as far as the horizon. They therefore need a number of "repeater masts" to cover even quite small countries.

Most countries are also linked by cables submerged under the sea. These still carry about half of the telecommunications traffic across the oceans, the other half going by satellite. Satellites have the advantage of being able to deal with all types of telecommunications. They can separately and simultaneously relay telephone, telex, computer, and fax messages, while receiving and re-transmitting high quality stereo radio signals and full color television pictures.

Short waves

Shortwave radio signals bounce up and down repeatedly between the Earth's surface and the ionosphere as they travel around the world.

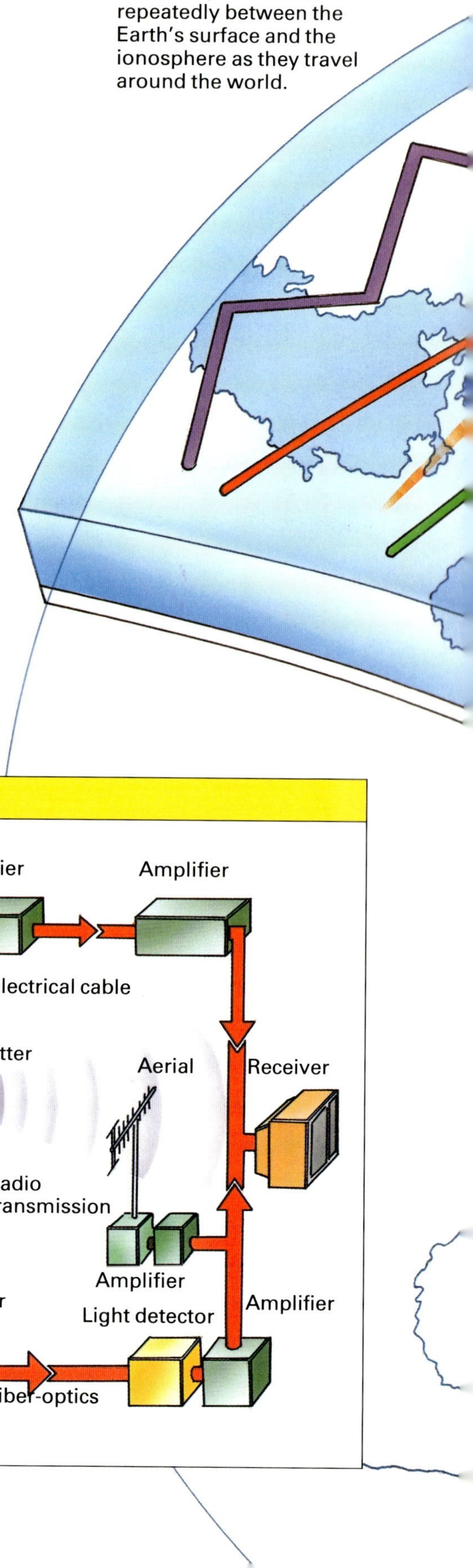

CONVERTING SIGNALS

A signal from a television studio can be sent to receivers in three ways. One is to transmit it as a radio wave. At the transmitter, a high frequency carrier wave is added. Receivers in the home pick up the signals using an aerial and convert them back into sound and pictures. Secondly, signals are sent via underground metal cables. Finally, the newest form of cable transmission uses fiber-optics instead of metal wires.

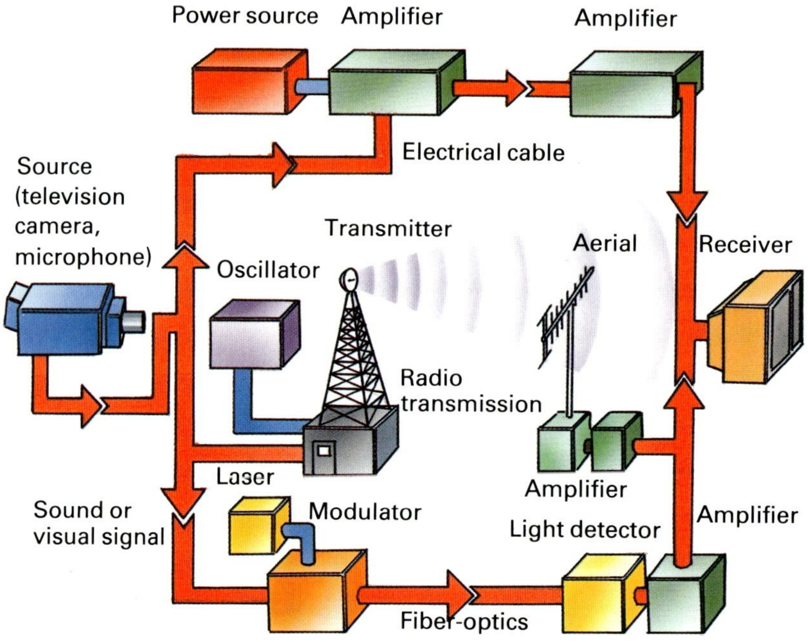

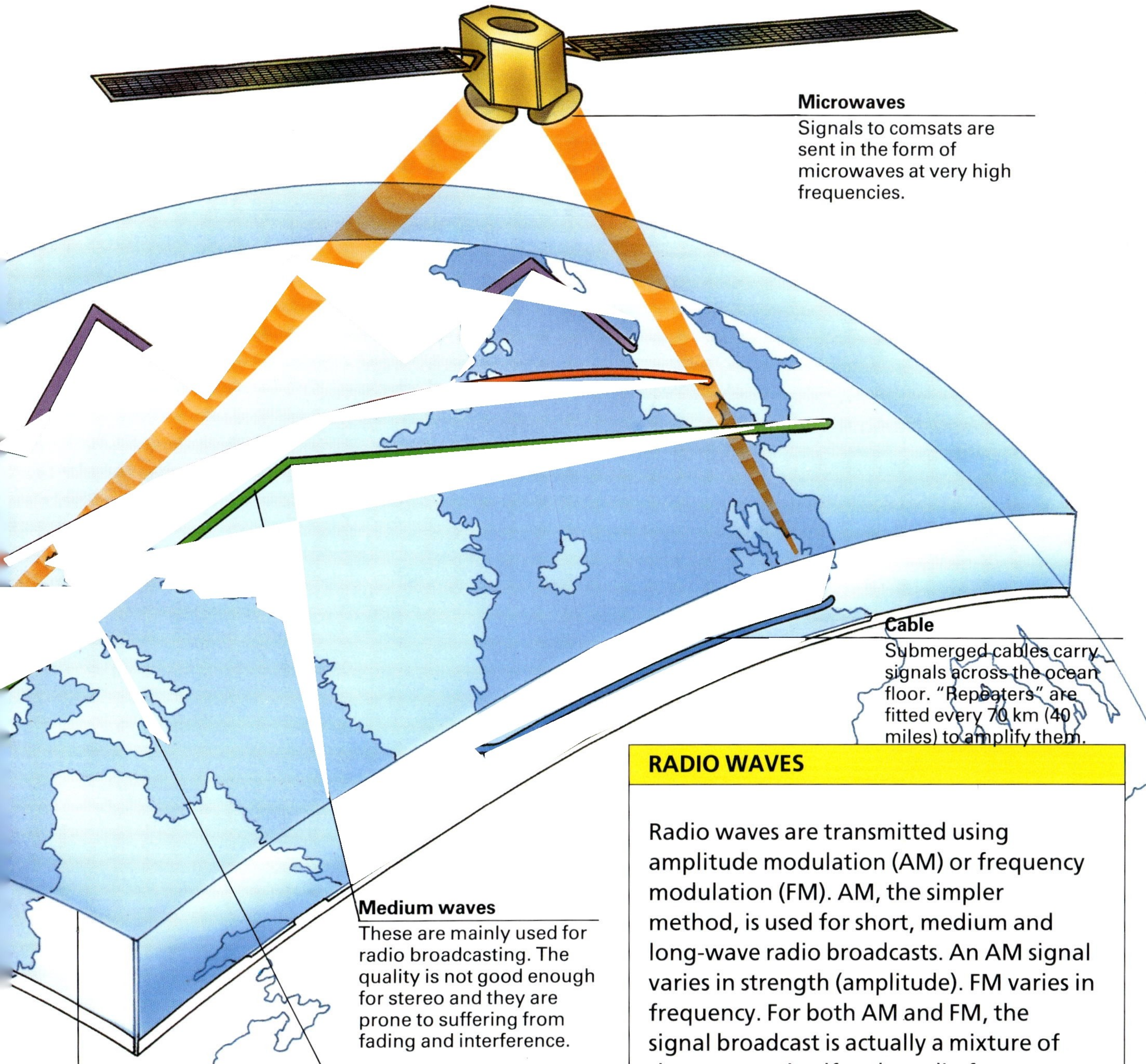

Microwaves

Signals to comsats are sent in the form of microwaves at very high frequencies.

Cable

Submerged cables carry signals across the ocean floor. "Repeaters" are fitted every 70 km (40 miles) to amplify them.

Medium waves

These are mainly used for radio broadcasting. The quality is not good enough for stereo and they are prone to suffering from fading and interference.

Ionosphere

The ionosphere is a layer of the Earth's atmosphere extending from about 60 to 1,000 km (35-600 miles) above the surface.

Long waves

Long waves can travel great distances because the signals follow the Earth's curved surface. However, the signals get considerably weaker the farther they travel, so reception can be poor.

The frequency of radio waves is measured in hertz (Hz). Radio broadcasts on the medium (or AM) bands are typically around 530 to 1,600 kilohertz (KHz). VHF (or FM) bands are much higher, 85 to 110 megahertz (MHz). The microwaves used by satellites are even higher frequency, measured in Gigahertz (GHz).

RADIO WAVES

Radio waves are transmitted using amplitude modulation (AM) or frequency modulation (FM). AM, the simpler method, is used for short, medium and long-wave radio broadcasts. An AM signal varies in strength (amplitude). FM varies in frequency. For both AM and FM, the signal broadcast is actually a mixture of the program itself and a radio frequency (RF) carrier wave.

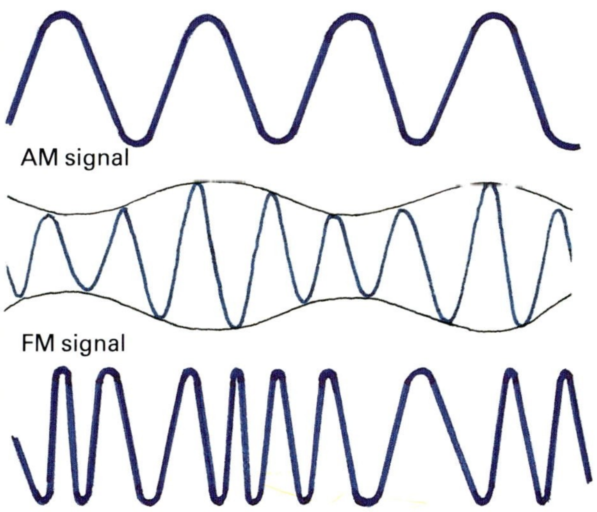

MODULE SAT

Building satellites is a growing industry. Aerospace companies manufacture a range of satellite parts called modules, which can be assembled in different ways. Some communications satellites, especially the largest ones, can deal with all types of signals and many types of receiving dishes. Other satellites are specially tailored to meet individual requirements, perhaps for communicating with ships at sea or aircraft in flight.

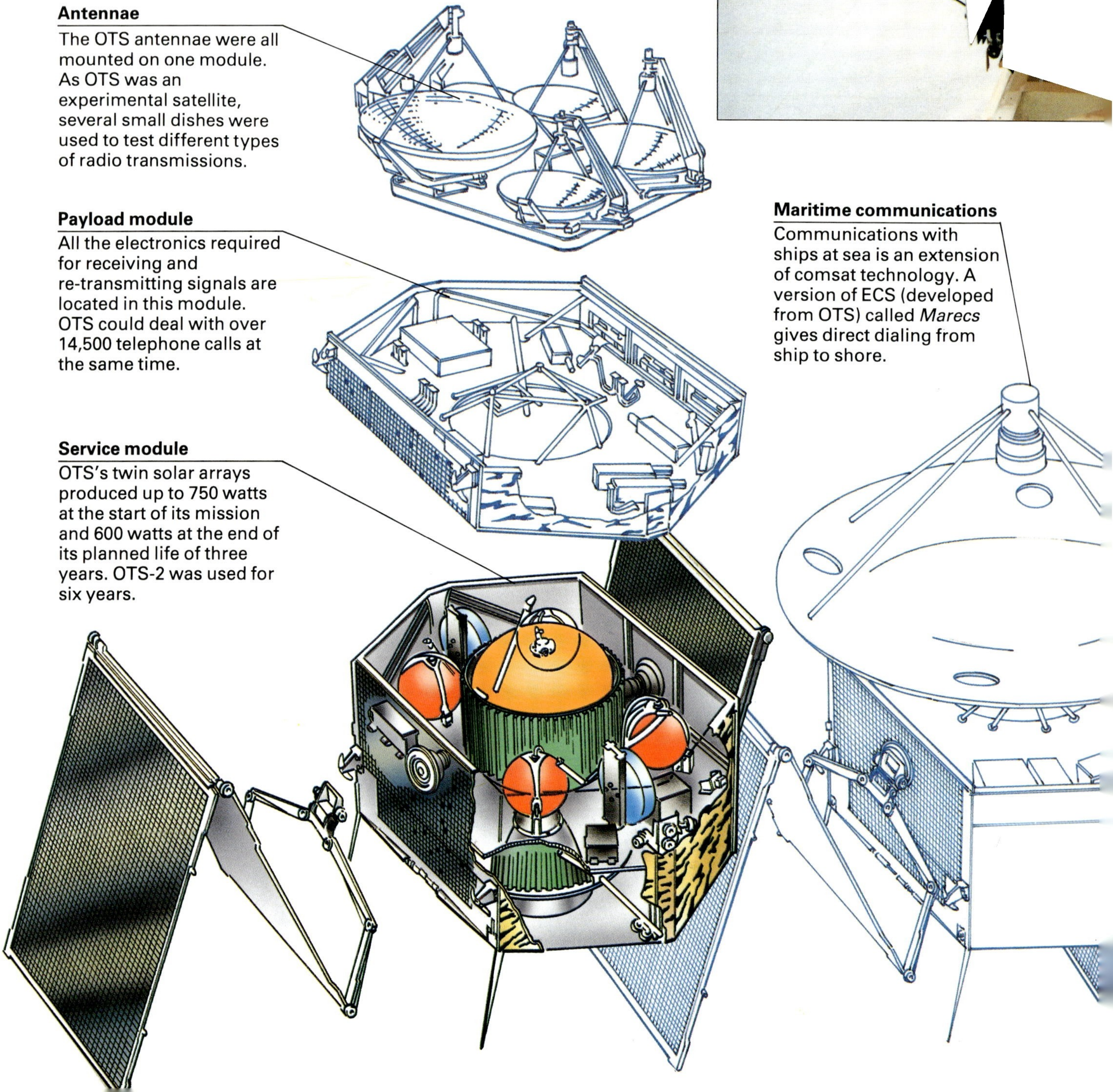

Antennae
The OTS antennae were all mounted on one module. As OTS was an experimental satellite, several small dishes were used to test different types of radio transmissions.

Payload module
All the electronics required for receiving and re-transmitting signals are located in this module. OTS could deal with over 14,500 telephone calls at the same time.

Service module
OTS's twin solar arrays produced up to 750 watts at the start of its mission and 600 watts at the end of its planned life of three years. OTS-2 was used for six years.

Maritime communications
Communications with ships at sea is an extension of comsat technology. A version of ECS (developed from OTS) called *Marecs* gives direct dialing from ship to shore.

The main body of a satellite is called the bus, which consists of several sections. For satellites that are designed to be built in various ways, the largest section of the bus, the service module, will probably be much the same for all versions. The service module consists mainly of the apogee motor that propels the satellite into the correct orbit, and its propulsion tanks. Tanks are also fitted here for the thrusters that position the satellite precisely, and the small thrusters themselves. The solar panels will also attach to the bus at the service module. The payload module carries the electronics and so this varies from type to type.

The European Space Agency's *Orbital Test Satellite* (OTS), launched in 1978, was a successful design for a modular comsat. The original craft components are shown in the diagram far left. The *European Communications Satellite* (ECS) was based on the lessons learnt from OTS.

A comsat bus during its construction. To prevent dust damaging the craft, this area is kept extremely clean.

Regional communications

By carefully shaping the antennae, narrow "spot" beams can be used for communications within a small area on the ground, where it may be difficult to use radio masts.

Aeronautical communications

Even nowadays aircraft can be out of touch with ground stations for long periods. A version of OTS for communications with aircraft was planned, but tests used *Marecs* instead.

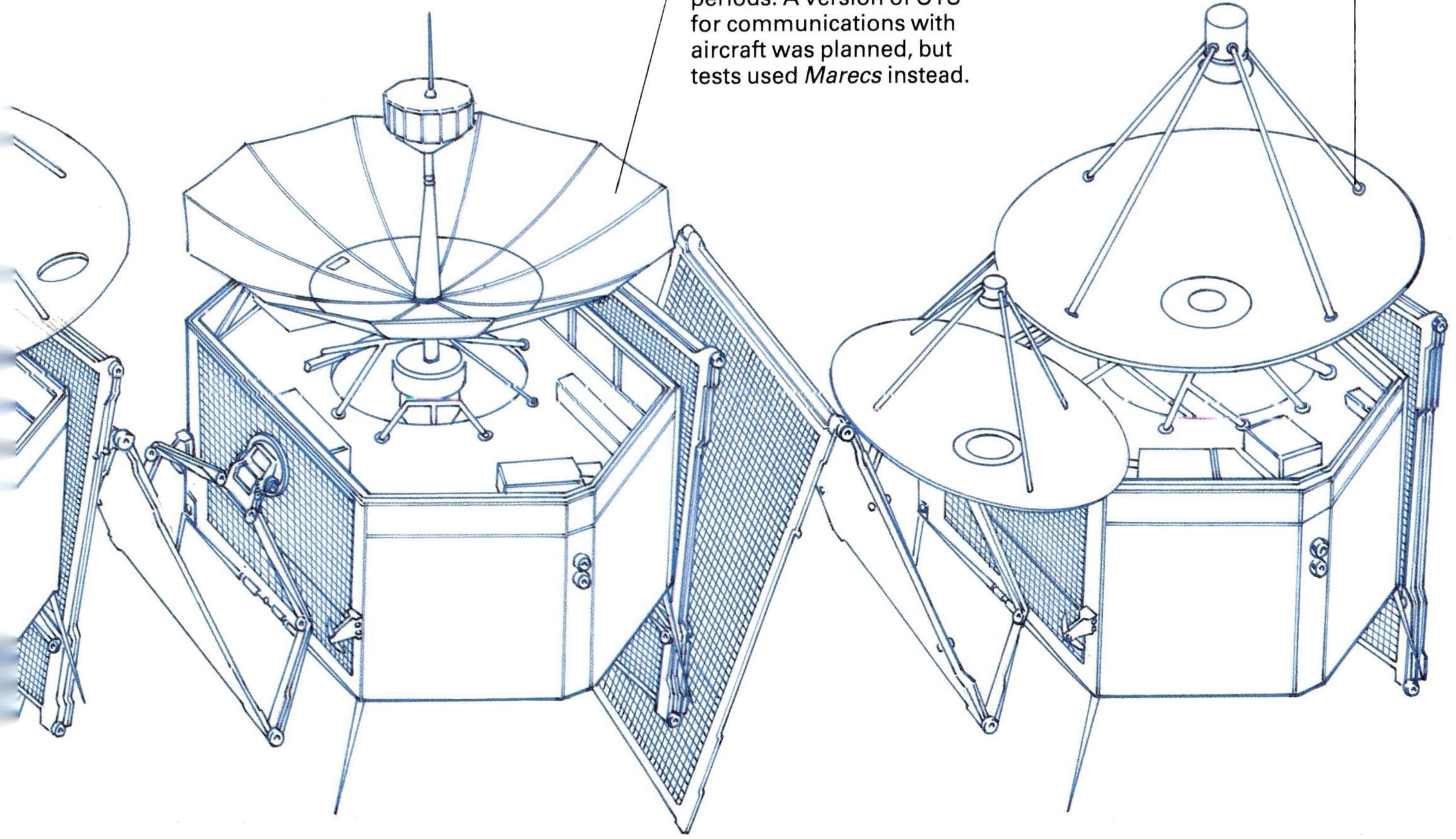

ON GEOSTATION

By far the majority of communications satellites are in a circular path around the Earth called a geostationary or synchronous orbit. The satellites circle the Earth at exactly the same speed as the Earth's rotation. The satellite therefore appears to stay over the same point on the Earth's surface. Here, it can be used as a relay station for radio transmissions 24 hours every day.

This idea seems obvious, but like so many ideas it was first thought of by only one person. In this case it was the British scientist and science-fiction writer, Arthur C. Clarke. In 1945 he wrote an article for the magazine *Wireless World*, called "Extraterrestrial Relays." This described how a satellite could be placed at a height of 36,000 km (22,000 miles) above the Earth's surface and used to relay radio signals. The article explained that just three satellites would be needed to cover the entire Earth. Today there are many, many communications satellites in geostationary orbit around the Earth.

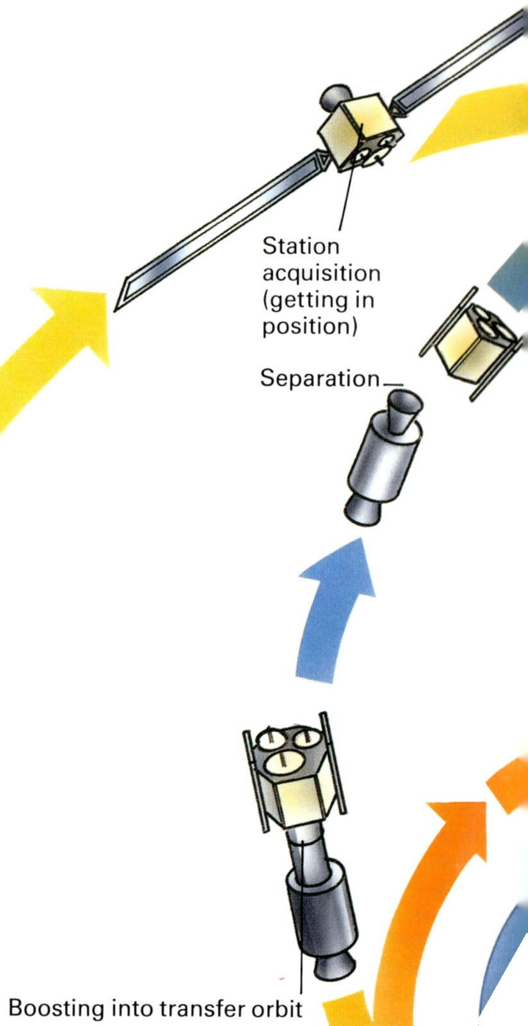

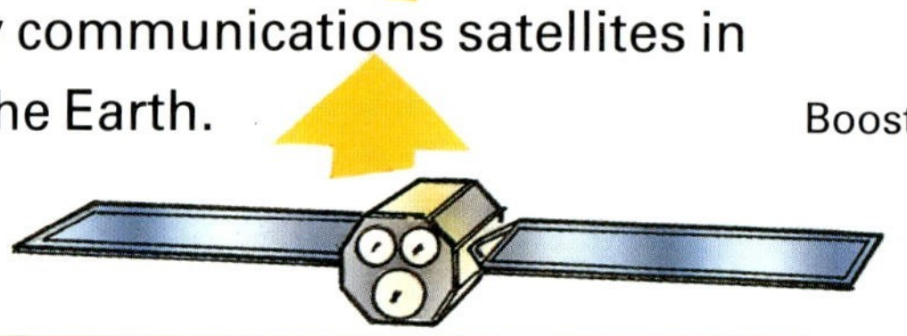

GEOSTATIONARY ORBIT

A communications satellite in a geostationary orbit is of course not stationary. It is actually traveling at 11,000 km/h (6,500 mph), which exactly matches the speed of the Earth rotating 36,000 km (22,000 mi) below. The satellite appears to be stationary above an observer on the Earth. Geostationary satellites must orbit around the equator. Observers north of the equator must look south for a comsat.

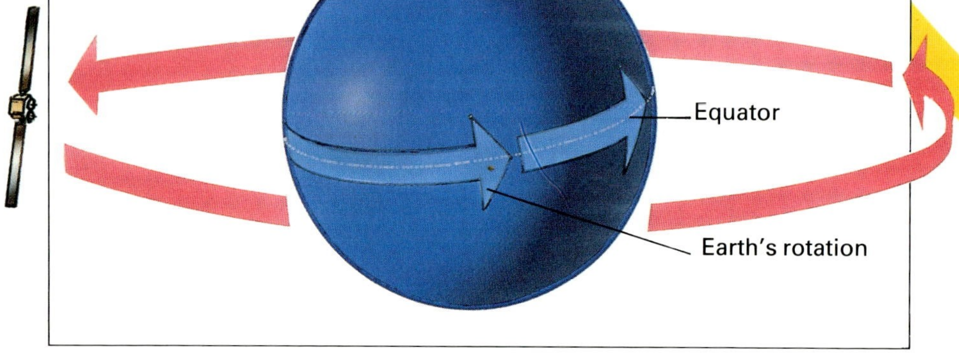

3. Geostationary orbit

A communications satellite will usually stay in the transfer orbit for two or three circuits while further tests are carried out on its on-board systems. When the satellite is next at apogee, its farthest point from the Earth, the satellite is turned to face the orbital path and its apogee motor is fired to keep it in the geostationary orbit.

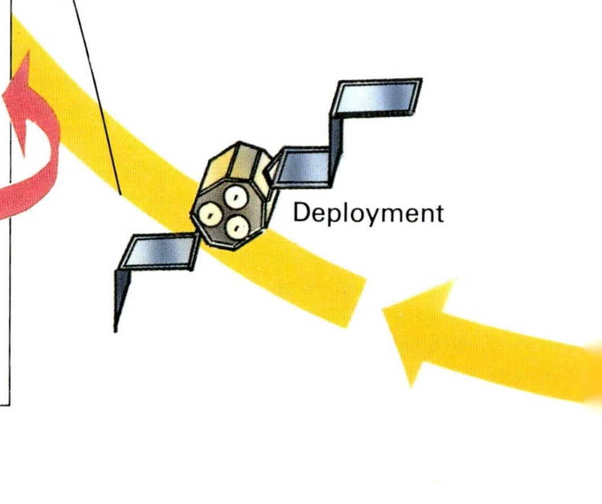

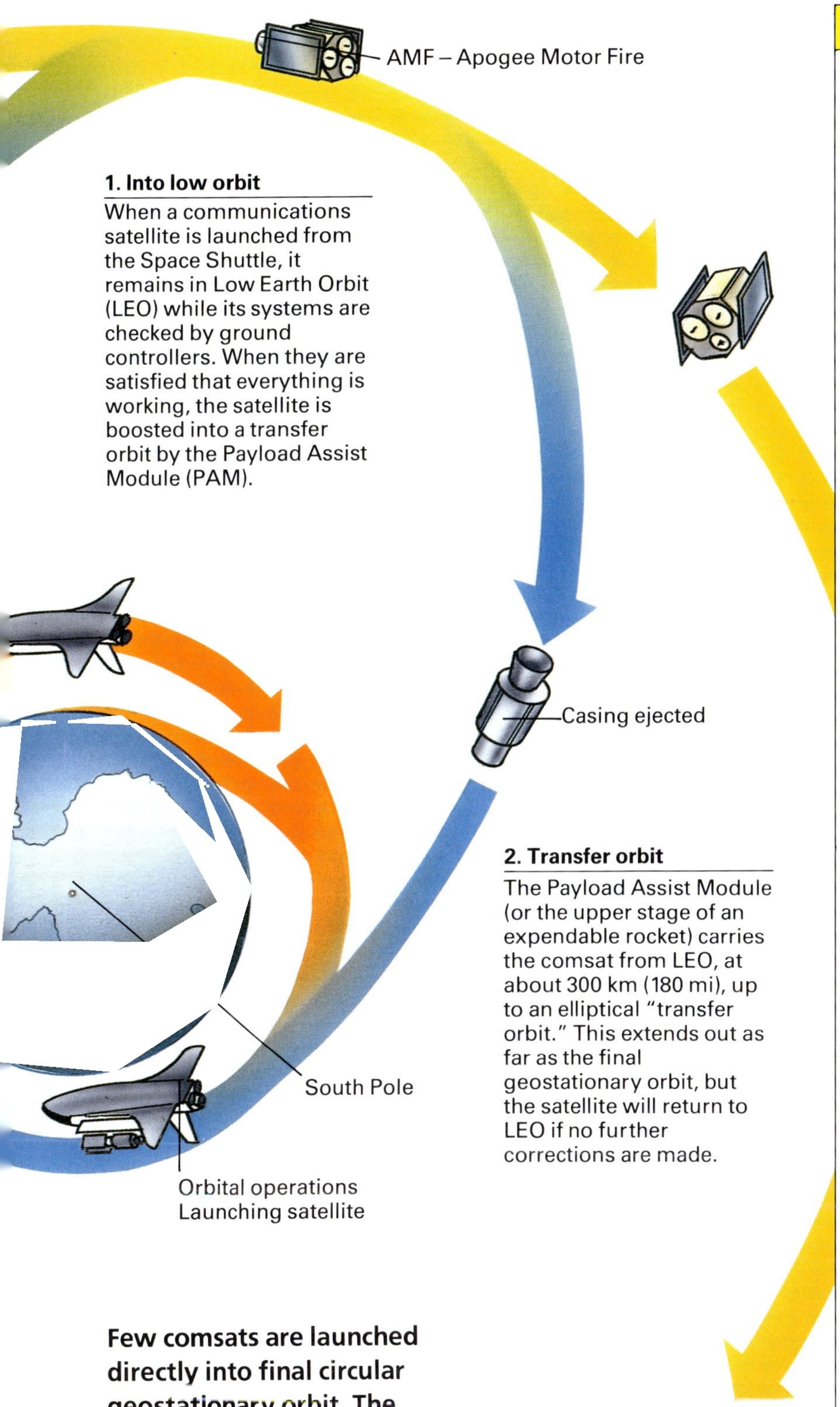

1. Into low orbit

When a communications satellite is launched from the Space Shuttle, it remains in Low Earth Orbit (LEO) while its systems are checked by ground controllers. When they are satisfied that everything is working, the satellite is boosted into a transfer orbit by the Payload Assist Module (PAM).

2. Transfer orbit

The Payload Assist Module (or the upper stage of an expendable rocket) carries the comsat from LEO, at about 300 km (180 mi), up to an elliptical "transfer orbit." This extends out as far as the final geostationary orbit, but the satellite will return to LEO if no further corrections are made.

Few comsats are launched directly into final circular geostationary orbit. The launching requires several stages, which are basically the same for a launch from the Space Shuttle or by expendable rocket. Since the loss of the Shuttle *Challenger* in 1986, there has been a rapid growth in the use of expendable rockets such as *Ariane*, *Proton*, and *Atlas*.

FOOTPRINTS

The original plan for orbiting "relay stations" by Arthur C. Clarke required just three satellites to send a message around the whole Earth. However, today most countries and satellite operators need to send many different messages targeted at particular areas of the Earth. The simplest beam shape to transmit from a satellite is circular, but no country is circular. So, the beam is shaped electronically to fit the target area, called the "footprint." Some areas, like the United States, are far easier to cover with a shaped footprint than, say, Europe with its many smaller individual countries with complex border shapes.

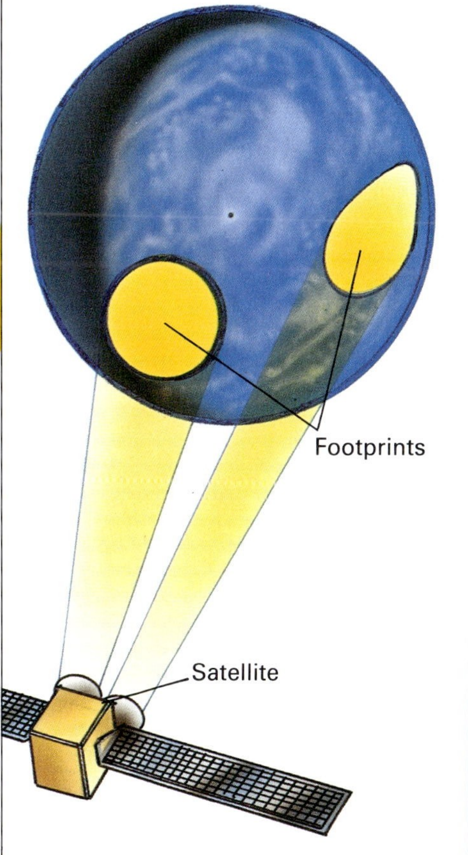

DEPLOYMENT

Since *Sputnik 1*, the first artificial satellite, the majority of satellites have been put into orbit by unmanned rockets. Nowadays, these rockets are also called ELVs, Expendable Launch Vehicles, to distinguish them from manned Space Shuttles. The American Space Shuttle was intended to be used for commercial satellites, and it did launch many comsats successfully. However, since the accident in 1986 that destroyed the Shuttle *Challenger* and its crew of seven, it was decided that the Shuttle would concentrate on exploration and science payloads. Commercial comsat launches have been moved back to ELVs. It is not known if the Soviet Union's Space Shuttles will carry comsats, though this is likely.

The European Space Agency's rocket, *Ariane*, launches most of the world's commercial comsats at the present time. There are several different versions of Ariane for launching a range of different payloads into orbit. However, the Soviet Union has the world's largest supply of ELVs, especially the *Soyuz* rocket. Japan, China, and India also have their own launch rockets.

The American Space Shuttle has launched several comsats. The Shuttle did the job of the first two stages of an ELV, lifting the satellite into Low Earth Orbit (LEO). From there, an additional stage, the Payload Assist Module (PAM), acted as the ELV's third stage, taking the comsat to geostationary orbit.

Modern comsats are large and, with their antennae extended, an awkward shape. So, for launch they have to be folded into a convenient shape to fit inside the ELV nose or the Shuttle payload bay. Drum shaped "spin stabilized" comsats, like *Intelsat 6*, increase their solar cell area by extending a second drum of cells from inside the first.

KEY
1. Launch configuration
2. Omni antenna in transfer orbit position
3. Solar panel extended
4. Omni antenna in final position
5. Transmitter reflector deployed
6. Transmitter reflector (and receiver reflector) in flight position
7. K-band antennae operational

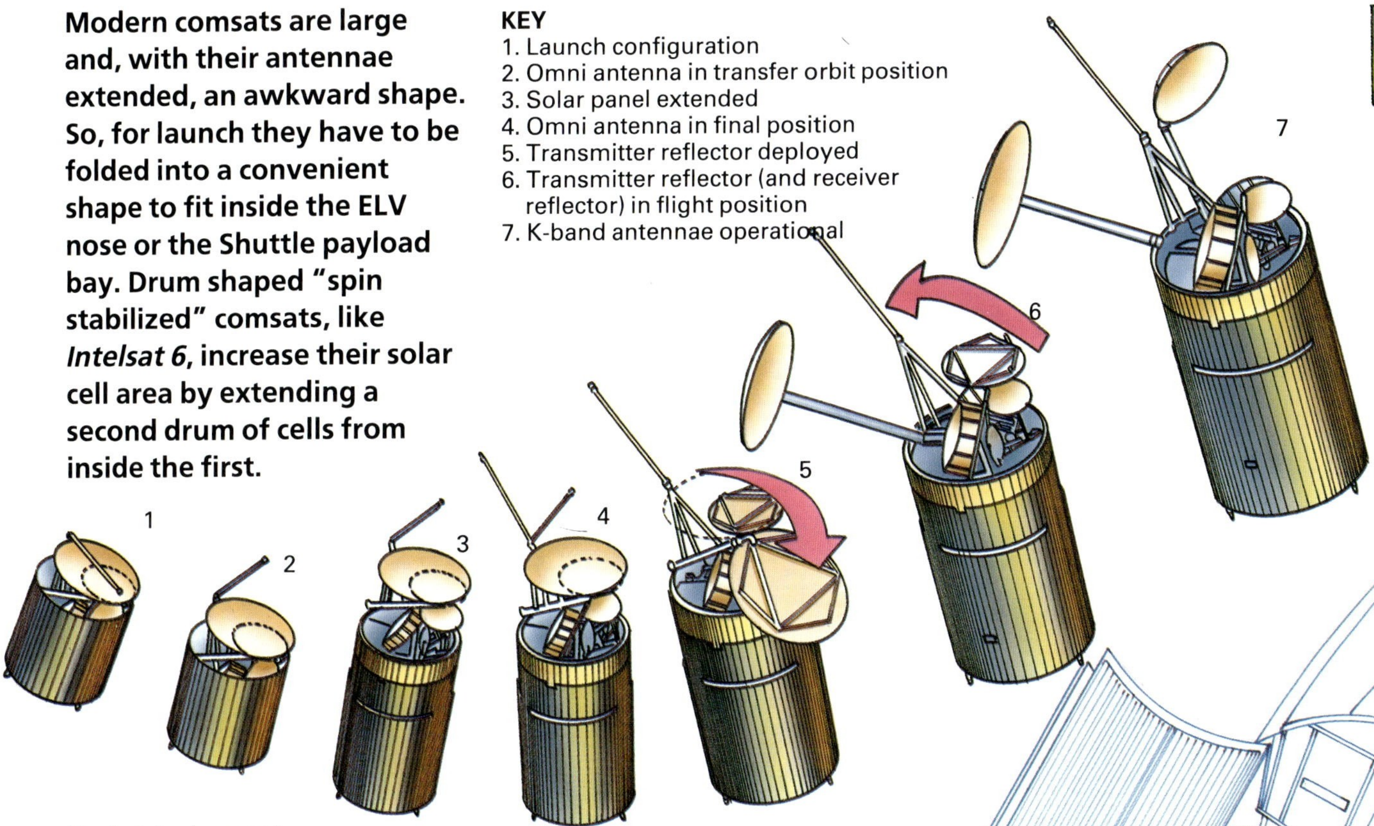

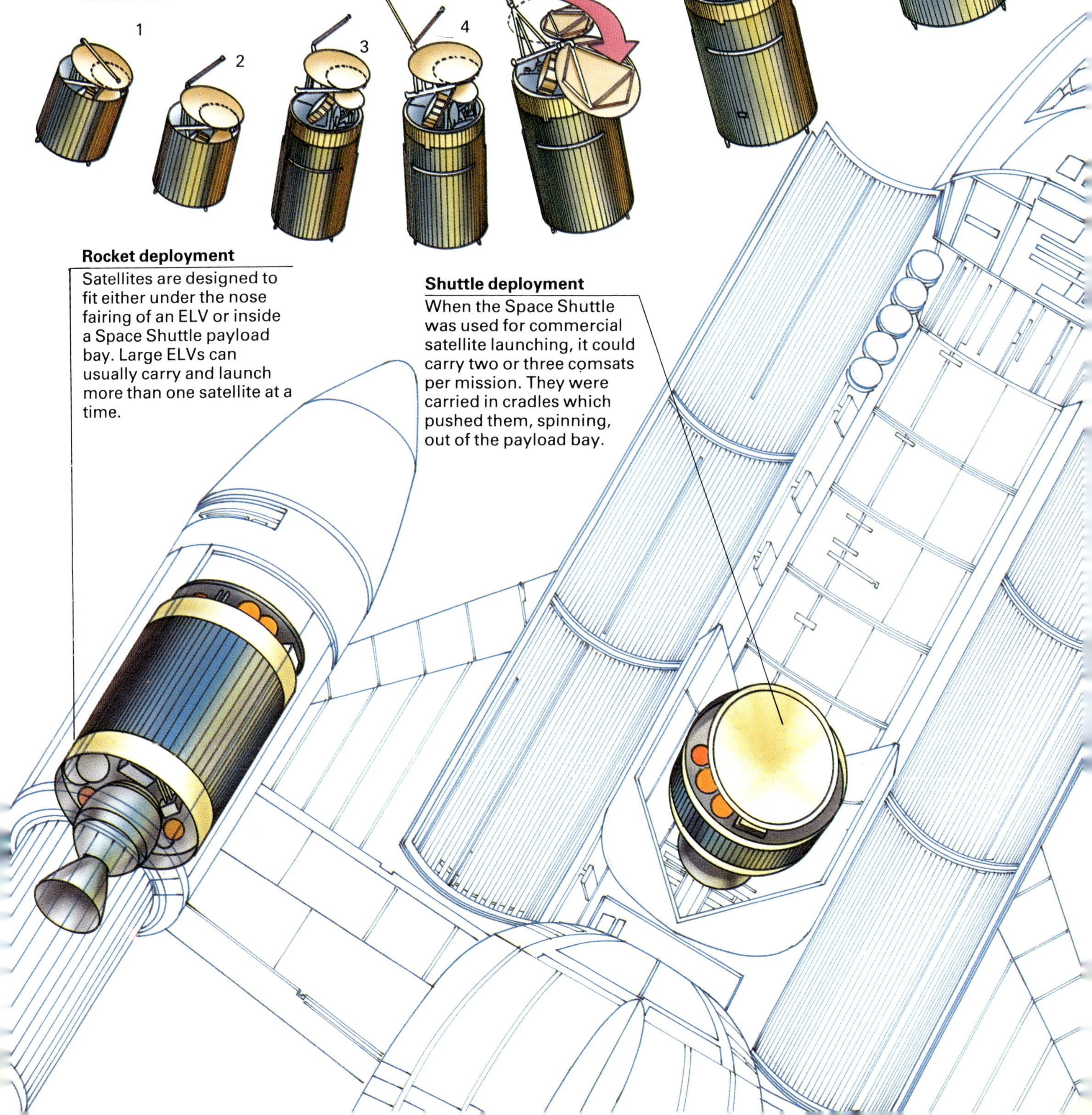

Rocket deployment

Satellites are designed to fit either under the nose fairing of an ELV or inside a Space Shuttle payload bay. Large ELVs can usually carry and launch more than one satellite at a time.

Shuttle deployment

When the Space Shuttle was used for commercial satellite launching, it could carry two or three comsats per mission. They were carried in cradles which pushed them, spinning, out of the payload bay.

STAYING IN ORBIT

Telstar was the first experimental comsat designed to receive TV transmissions and re-transmit them back to Earth. On test, giant dish-shaped antennae on Earth had to move to track it across the sky. This was because *Telstar* circled the Earth in a Low Earth Orbit. It moved rapidly from one horizon across the sky to the other horizon. The advantage of the geostationary orbit is that comsats placed there appear to be stationary in the sky, constantly relaying messages from and to stationary dish antennae on the ground.

Although comsats appear to stay in the same place in their orbit, external forces can make them wander slightly. Small rocket thrusters mounted at the corners of the satellite's body are used to correct these movements. Satellite operators guarantee that their satellites will stay in position to an accuracy of 0.1°. This allows manufacturers to design dish antennae capable of receiving signals from a satellite within this tiny range of positions. But even 0.1° over 36,000 km (22,000 mi) is a 63 km (38 mi) error in the comsat's position.

A comsat orbiting the Earth has to rotate around itself so as to keep its antenna dishes pointing toward the ground stations. This slowly spinning movement is started when the craft first reaches orbit. From time to time, small errors in the these movements have to be corrected.

Sensing the Sun

As the main body of the comsat rotates to stay aligned with the Earth, its solar panels also have to rotate to face the Sun. This is vital to generate the maximum electrical power. Each panel is driven by an electric motor and steered by a Sun Sensor that can monitor the comsat's position relative to the Sun.

DISTURBANCES IN ORBIT

The external influences on a satellite are many and varied. Although the Earth is always thought of as round, it is not actually a perfect sphere. The southern hemisphere is larger than the northern hemisphere. The Earth's center of mass is therefore south of the Equator. The Earth's gravity, acting from this point, tends to pull the satellite out of its orbit around the Equator.

The Moon's gravity also influences a satellite's position. In the same way as the Moon causes tides on Earth, it also makes a comsat repeatedly move toward and away from the Earth.

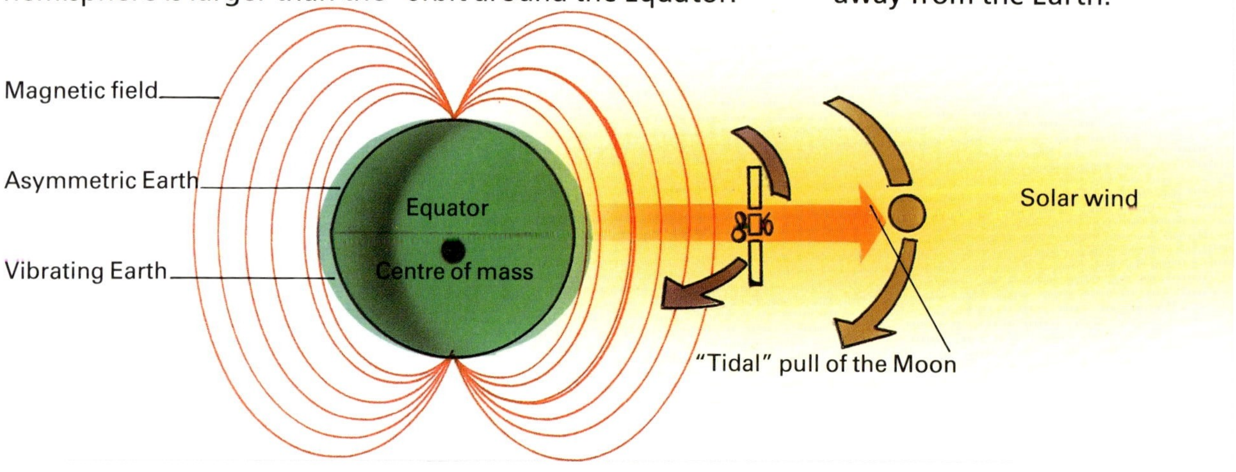

Reaction wheels

Reaction wheels are small fast-spinning flywheels used to keep the craft stable. A modern comsat uses three or more, one on each axis of the comsat's body. Sensors detect the craft's movements and instruct the appropriate wheel to speed up or slow down, moving the craft in the opposite direction.

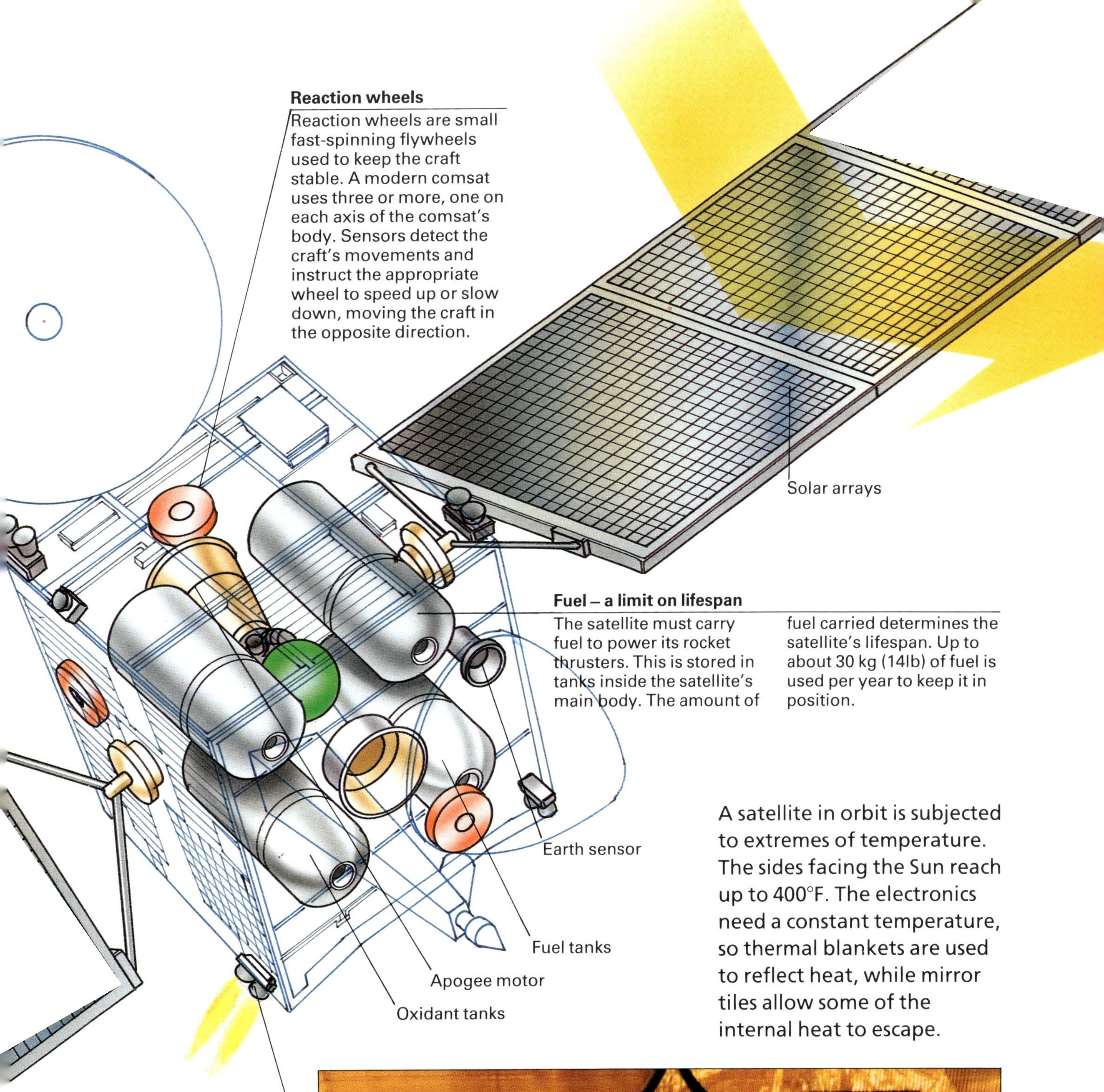

Fuel – a limit on lifespan

The satellite must carry fuel to power its rocket thrusters. This is stored in tanks inside the satellite's main body. The amount of fuel carried determines the satellite's lifespan. Up to about 30 kg (14lb) of fuel is used per year to keep it in position.

A satellite in orbit is subjected to extremes of temperature. The sides facing the Sun reach up to 400°F. The electronics need a constant temperature, so thermal blankets are used to reflect heat, while mirror tiles allow some of the internal heat to escape.

Thrusters

Thrusters are small rocket motors fitted to the corners of the comsat's body. They are not very powerful, giving less than a pound of thrust, but this is enough to move the craft in the vacuum of space. To twist the craft, one set of thrusters fires in a short burst. The satellite keeps rotating until the opposite set of thrusters is fired to stop it.

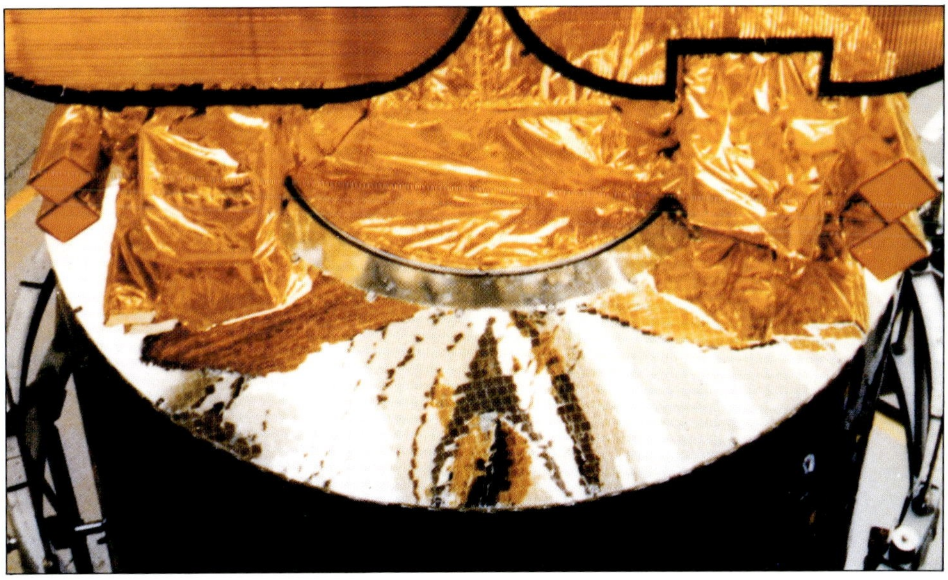

SUN POWER

A satellite in orbit has to be supplied with electrical power for its transmitter and receivers, and other on-board systems such as thermal control and stabilization. All commercial communications satellites use solar cells to generate this power. Other methods of producing electricity, such as small nuclear power generators, could be used, but these tend to be restricted to specialist military craft and planetary probes. It is simpler to generate power from solar cells and of course the source of the power, sunlight, is free.

The solar cells can be arranged in two main ways on a comsat. In three-axis stabilized craft, like *Olympus* or *Telecom 1*, the solar cells are fixed to panels that extend from opposite sides of the satellite's body like aircraft wings. On spin-stabilized satellites, like the Hughes *HS 376* design or *Intelsat 6*, the solar cells are positioned all over the cylindrical body of the craft. In this design, fewer cells face the Sun at any one time, but the cell arrays do not need to be steered to face the Sun as cells on the wing-like panels do.

Solar cells are not very efficient at converting sunlight into electricity. Only about a tenth of the light's energy is converted into usable power. The main advantage is that the source of energy is free and continuous.

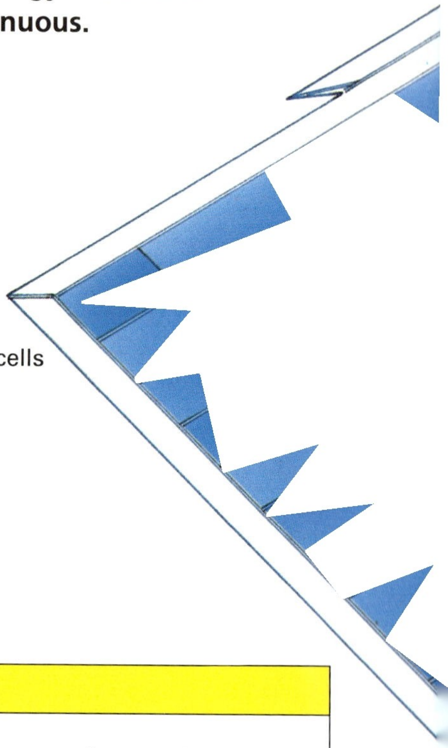

STORING POWER FOR THE ECLIPSE

Because the Earth's axis is tilted to the plane of the Solar System and comsats orbit the Earth's Equator, a comsat's solar panels "see" the Sun for most of the year. However, twice a year, in spring and fall, the satellite enters the Earth's shadow. The solar cells are deprived of light and cease to produce power. On-board batteries, charged by the cells, keep the craft going until the eclipse has passed.

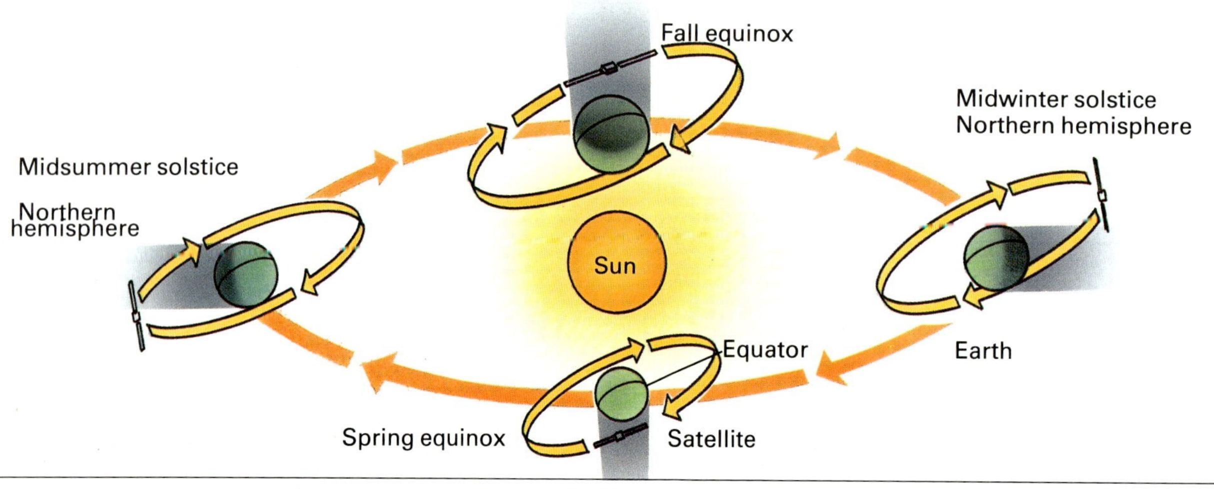

ALWAYS FACING THE SUN

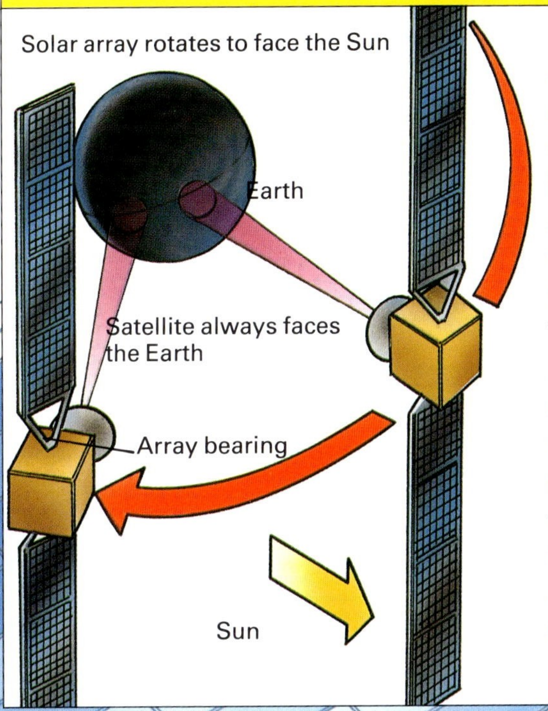

As a three-axis stabilized comsat orbits the Earth, the dish antennae on its main body have to face the Earth at all times. The solar arrays must face a different object, the Sun, to generate electricity. The arrays are positioned along a north-south line, where they have the least obstructed view of the Sun. Rotator motors constantly turn them so that they always face the Sun. Modern solar cell arrays can generate about 14 watts per square foot.

Creating electricity

Light consists of photons, individual "packets" of energy. They pass straight through a solar cell's outer layer and excite (or energize) electrons in a layer of silicon. The electron movements produce an electric current.

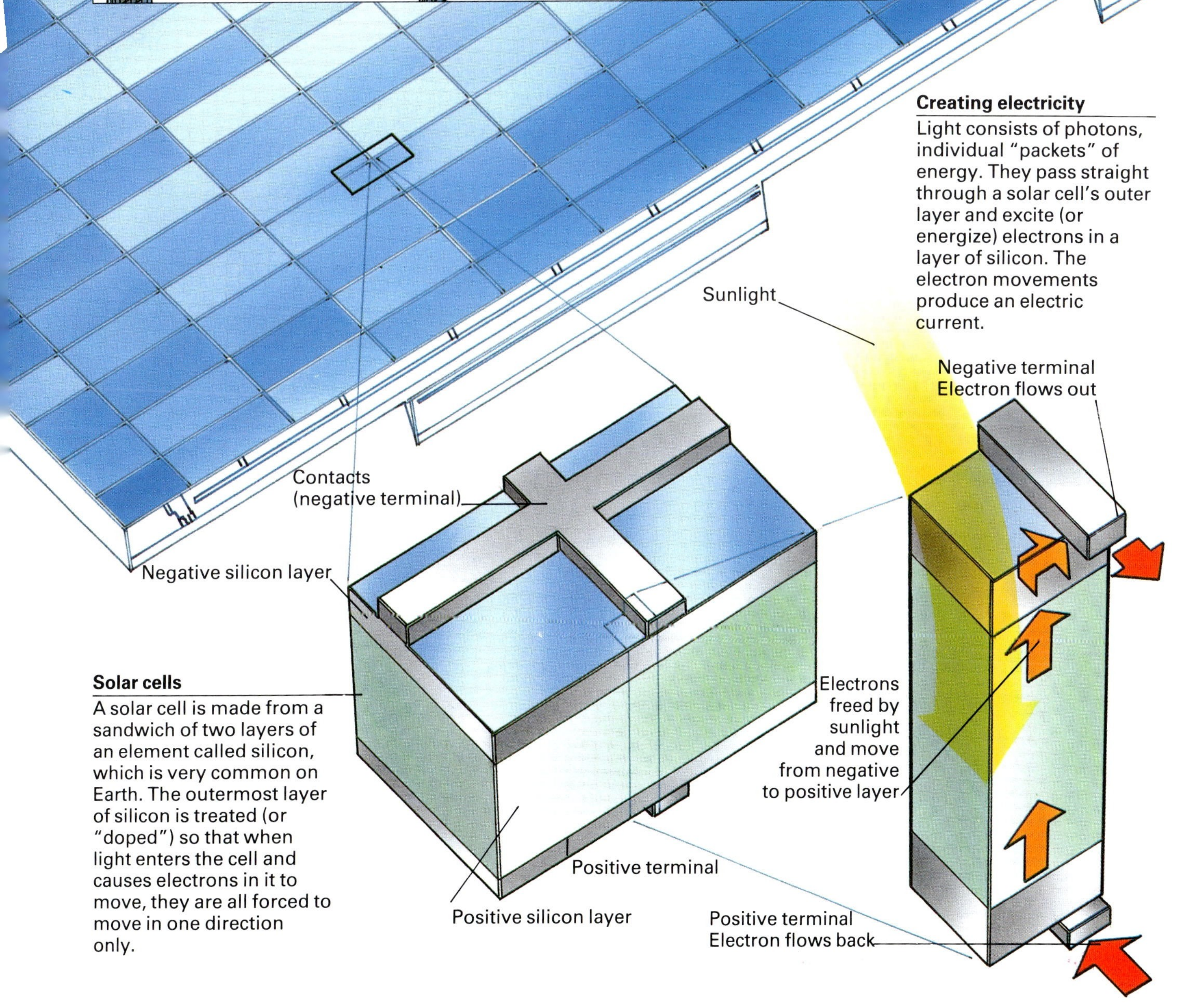

Solar cells

A solar cell is made from a sandwich of two layers of an element called silicon, which is very common on Earth. The outermost layer of silicon is treated (or "doped") so that when light enters the cell and causes electrons in it to move, they are all forced to move in one direction only.

COMSATS AT WORK

The earl ... ercial co ... s, like *Early Bird* an ... ould ... amount of transı ... individual specially-e...

Modern satellites have ... amount of traffic in the form o ... telex, and other electronic transmi... s, ... them to a variety of ground receivers. ... comsat's transmitting antennae is therefor ... problem for satellite engineers.

The signals are received and re-transmitted by components called "feedhorns" positioned in front of th large dishes. The dishes themselves collect the weak incoming signals, also called the uplink, and focus them in a narrow beam onto the feedhorn. The amplified signal re-transmitted back down to Earth, the downlink, is also directed by the antenna dish. The transmitting beam can be targeted onto a particular country or area by careful choice of the signal frequency, and of the dish shape and size.

Transmitter
Some comsats use the same dish antenna to receive and transmit all signals, but most comsats have several dishes.

Telemetry
In addition to communications signals, a comsat must relay its own control and data signals, called telemetry. These use their own antennae.

TRANSPONDERS

Comsats carry a number of transponders. A color television signal might have one transponder to itself, while another transponder carries hundreds of telephone calls. The largest comsats carry up to 50 transponders. When a signal is received by the satellite, it is very weak. A filter removes any background noise before the signal is amplified, converted to a new frequency and transmitted.

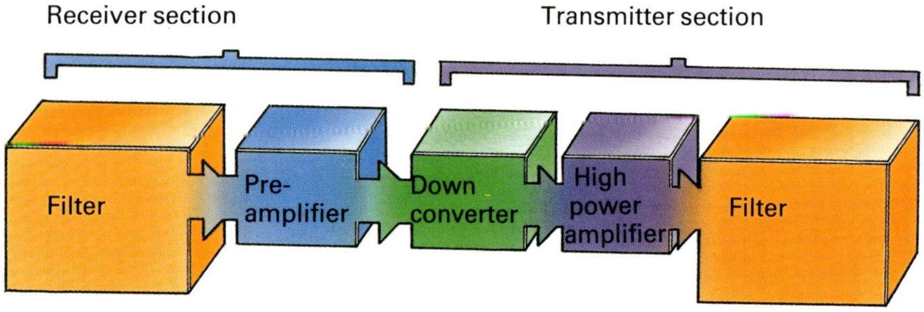

A dish shape is the result of a great deal of complex research and experimentation. By clever design, a dish may be able to receive one signal from its front surface while transmitting a second signal from its rear surface.

Transponders

The transponders are located inside the comsat's body. Even with microelectronics, they can be large devices. Their size may determine how many can be carried.

The number of transponders is related to the solar panel power output. If 1 kilowatt is available, this could run four 250-watt transponders or 10 100-watt transponders.

Receiver dish

The antenna dishes are usually not a simple circular shape. They are more commonly a complex curved shape that is determined by the offset position of the feedhorn.

FEEDHORNS

A feed-horn can be positioned in one of two ways. It may be fixed directly over the center of the dish. However, the feedhorn and its supports block some of the signal and so it is more common to position the feedhorn off to one side of the dish. This is called offset mounting.

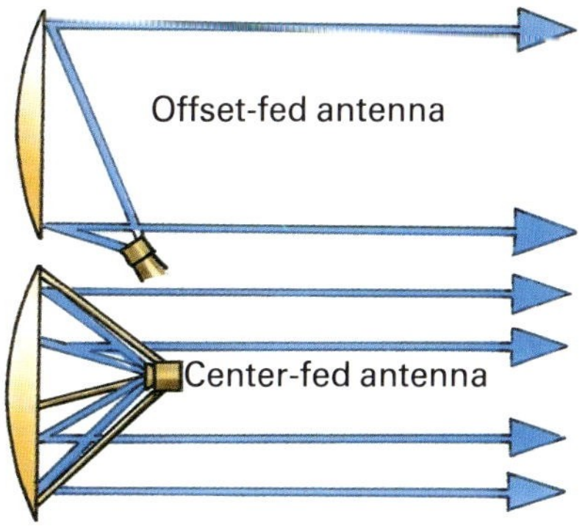

Britain's base for the first comsat trials was built at Goonhilly in Cornwall. It is still one of the world's largest ground stations.

As radio waves travel at the speed of light, it is possible to send a signal all the way around the Earth in less than a second by using three communications satellites.

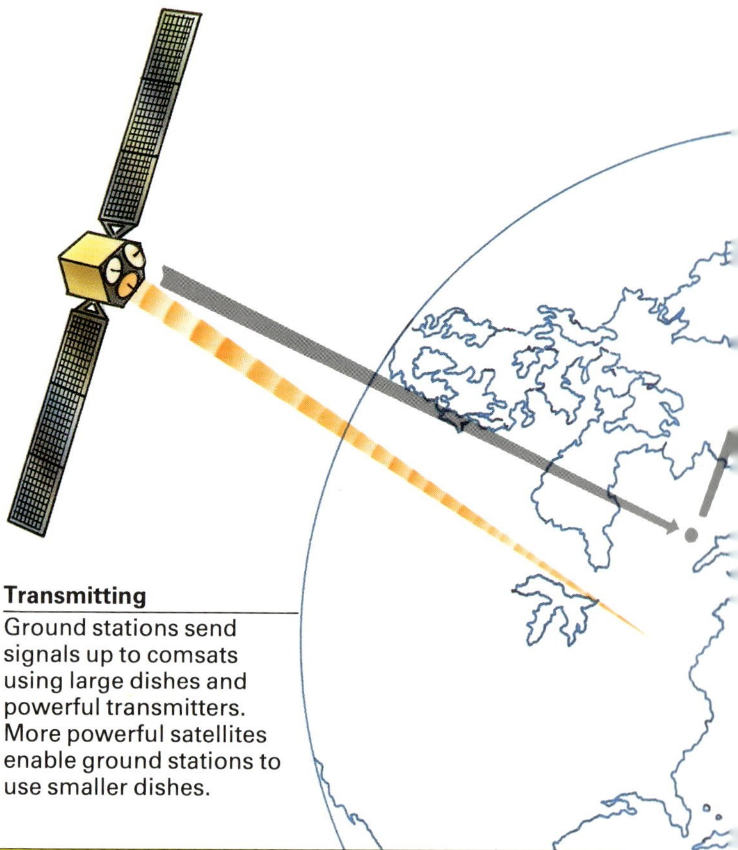

Transmitting

Ground stations send signals up to comsats using large dishes and powerful transmitters. More powerful satellites enable ground stations to use smaller dishes.

TRANSMITTING

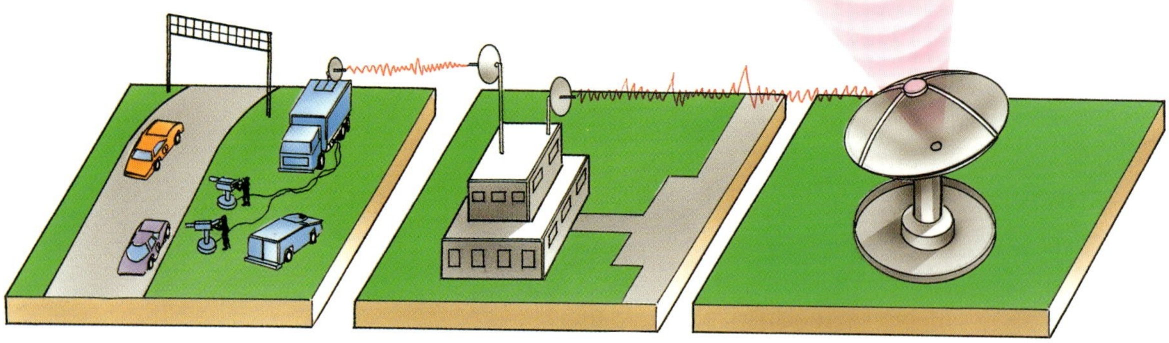

An event can be relayed in full color with stereo sound to anywhere in the world. The signals are controlled from a small studio and may be transmitted from a portable ground station.

At the broadcasting center, signals can be sent on without modification, but it is also possible to mix in additional pictures, sound or captions from another studio. Signals usually leave these centers by microwave.

When the signals arrive at the ground stations, they are converted to the satellite's uplink frequency and boosted by a high-power amplifier. Even so, the signals received by the satellite are very weak.

WIRE-LESS WORLD

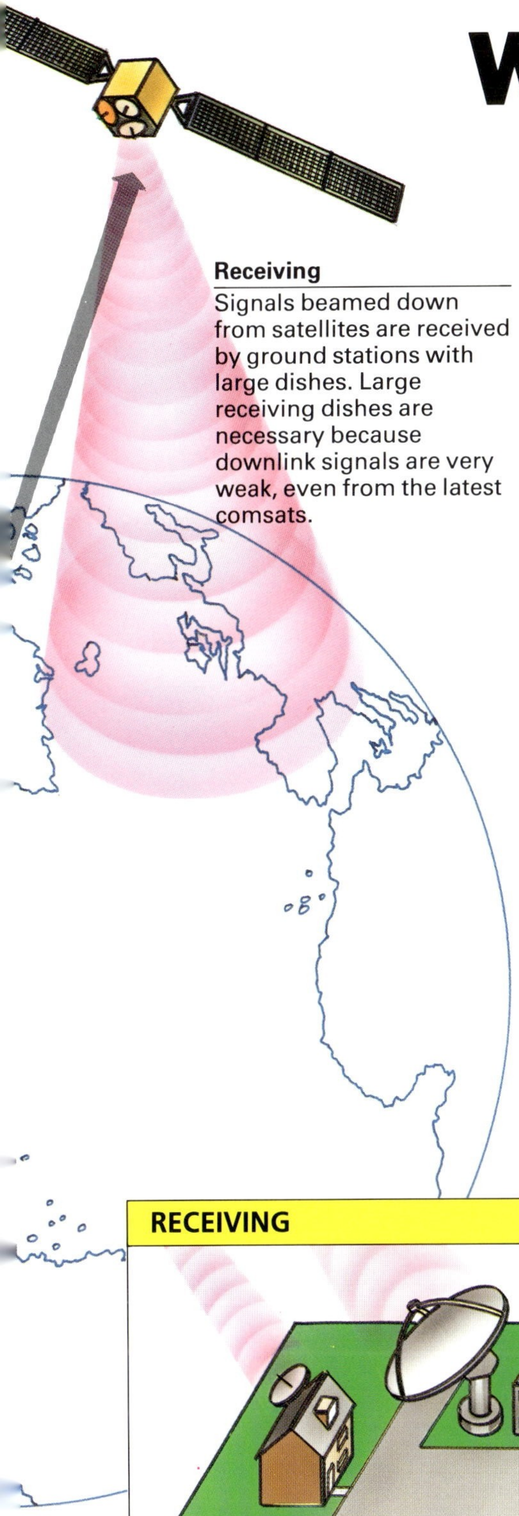

Receiving
Signals beamed down from satellites are received by ground stations with large dishes. Large receiving dishes are necessary because downlink signals are very weak, even from the latest comsats.

Comsats have revolutionized communications around the world. Although all the continents and most individual countries have been connected together electronically for many years, this had to be done by using short radio links or cables. Both methods suffer from problems, especially unwanted noise that can make messages difficult to receive.

The launch of the world's first communications satellites in the early 1960s immediately improved the situation. Any country with suitable equipment could join this growing worldwide network. The world's less developed countries have benefited from this capability of satellite communications. Before the Space Age, they had limited internal communications. Indonesia is a very good example. It consists of over 6,000 small islands. Its own comsats, called *Palapa*, now ensure that the whole island group is linked together electronically.

The latest use of high-power comsats is Direct Broadcast by Satellite (DBS) of television programs to roof dishes.

RECEIVING

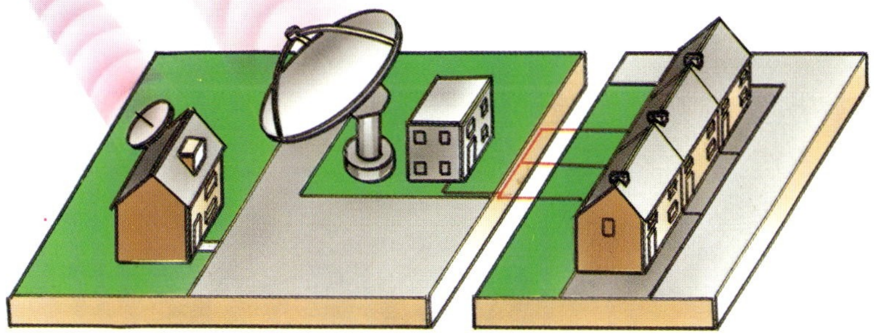

Until recent years, the signals sent back to Earth from a communications satellite were so weak that they could only be picked up by the large dishes 10 m (33 ft) across, used by commercial ground receiving stations.

The ground station would send the signals to TV viewers' homes from a transmitter or by cable. DBS uses more powerful satellites, whose signals can be picked up by small dishes fixed to the roofs of individual houses.

A RISKY BUSINESS

Launching a satellite is a risky business as so many things can go wrong. The launch rocket could fail during liftoff, or one of the upper stages fail while the rocket climbs toward orbit. The fairing containing the satellite could fail to open, or the satellite release mechanism might not operate. Even if the launch goes well, the satellite may still end up in the wrong orbit, or its on-board systems may not work. Satellite launches are so common now and failures are very rare. Catastrophes like the loss of the *Challenger* Shuttle due to a design fault are even rarer. To guard against failures in a comsat's on-board systems, duplicate systems are used. If one system fails, a "backup" takes over.

Repairs to manned spacecraft such as *Skylab*, *Salyut* or *Mir* can be done because crews are present on-board. Repairs to unmanned satellites are not so easy. The Space Shuttle can take crews to satellites such as *Solar Max* and repair them in LEO or return them to Earth for repair. The Space Shuttle brought the *Palapa B2* and *Westar 6* satellites back to Earth for repair.

Shuttle astronaut Joseph Allen used his Manned Maneuvering Unit (MMU) to "fly" across to the *Westar 6* satellite. Having stopped it spinning, he then used the MMU jets to return the satellite to the Shuttle's payload bay. There, astronaut Dale Gardner helped him to fix the satellite securely to a platform in the payload bay for its return to Earth.

The successful launch of a spacecraft depends on a multitude of mechanical and electrical parts working correctly. Every part is thoroughly tested, but faults can still occur. A rocket exploding on the launch-pad is extremely rare. Failures of the second or third stages are more common, usually due to an engine not firing. Once the satellite is released (called deployment) it may be struck by pieces of the discarded rocket stages or possibly by a meteor, though this is a very rare occurrence.

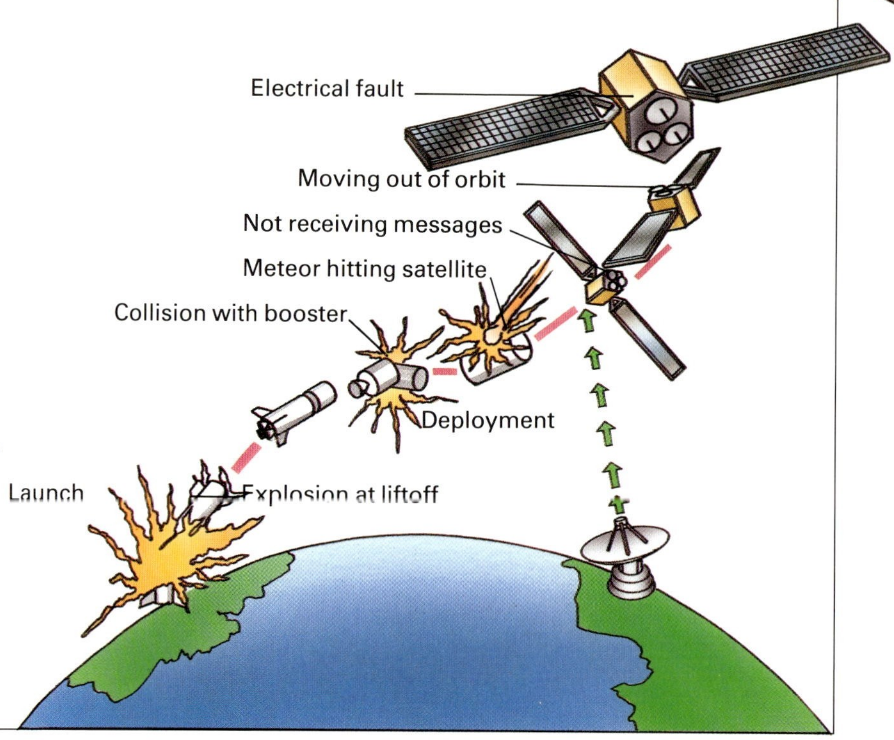

FOR SALE

THE FUTURE

Communications by satellite have advanced a long way since the first tests in the early 1960s. Then, it was an achievement just to send a fuzzy black-and-white television picture from one side of the Atlantic Ocean to the other. Now we regularly watch color television pictures relayed from all over the world via one or even two communications satellites. The satellites handle thousands of telephone calls at the same time. In future, communications satellites will link space stations with receiving stations on Earth. They will also continue to find new applications here on Earth. Miniaturized equipment for sending signals to communications satellites enables TV camera crews to carry their own satellite stations as "backpacks," enabling them to transmit live pictures from isolated locations anywhere in the world. Even smaller transmitters and receivers will enable individual people to communicate with each other. Using smaller, lower-power equipment on Earth will require bigger, more powerful satellites in orbit. They will need larger launch rockets, which are now being designed.

In the past comsats had to be small enough to fit inside the nose of a rocket or inside the Space Shuttle's payload bay. In the future, enormous communications satellites could be made with girder arms that "unfurl" in space (right). Manned space stations will also need their communication antennae, built in sections (below). The vast "antennae arms" imagined some years ago, dwarfing the Space Shuttle (below far right), may no longer be necessary as electronic circuits become smaller and more powerful.

Early next century, person-to-person communications by satellite may be possible using tiny wrist communicators. This may require large, powerful comsats and thousands of radio channels so everyone who wants to use a channel can do so.

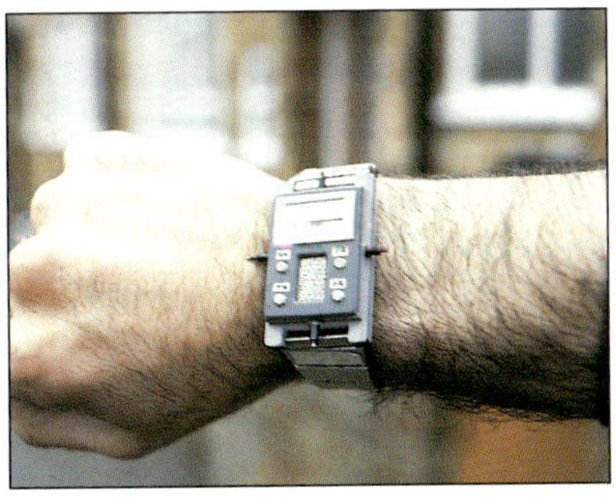

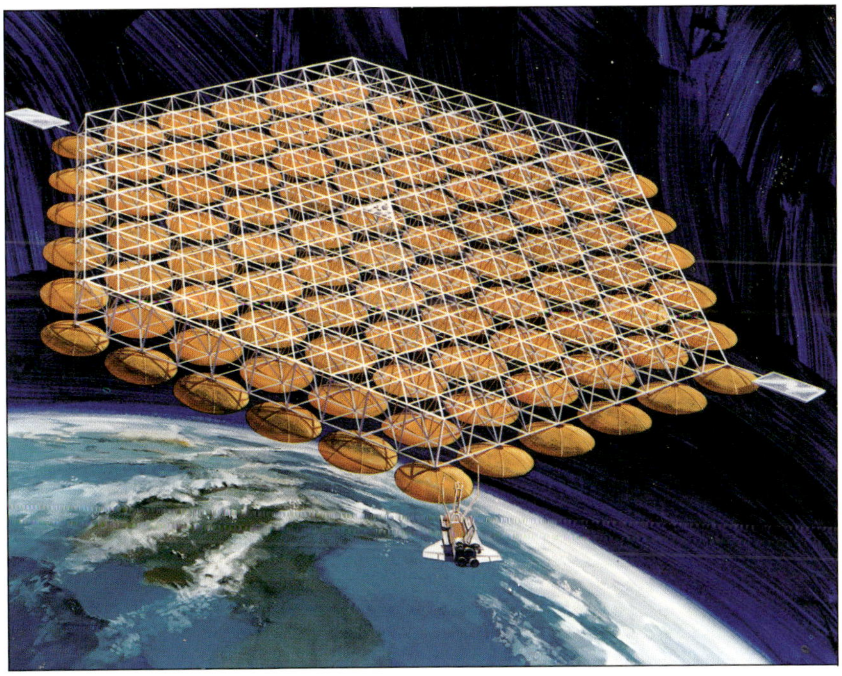

THE YOUNG ENGINEER

Experiments involving communications satellites are difficult to do at home. But with a bit of ingenuity you can use everyday household items to demonstrate some of the scientific principles and problems that engineers and scientists are faced with when they have to design and build full-size communications satellites.

In orbit

A satellite stays in orbit around the Earth because the outward force from its speed and course is exactly balanced by the force of gravity trying to pull it and the Earth together. The same effect can be demonstrated easily in the home by fixing a small ball or a bead to the end of a piece of string. When the ball is spun around in a circle, it pulls outward and so behaves like a satellite in orbit around the Earth. Swinging the ball around gives it "orbital speed." Its mass, however, pulls outward, away from the center of the orbit. The string acts like the pull of the Earth's gravity, preventing the ball's escape – unless you let it go!

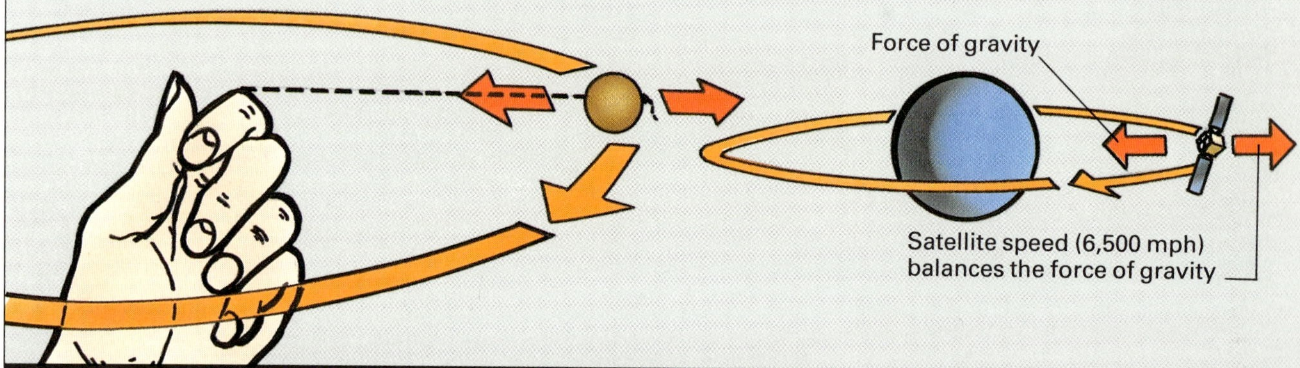

Maneuvering in orbit

Hang a cardboard box, representing the satellite's body, on a single thread, so that it spins easily. Put a balloon inside and bring its neck out through a hole cut in one corner of the box. Blow up the balloon and let the air escape. The box swings in the opposite direction, just like a real satellite using its thrusters. Movements in different directions can be made by cutting holes in different corners. Adding solar panels to the box and blowing them with a hair dryer or fan shows the effect of the solar wind on the satellite.

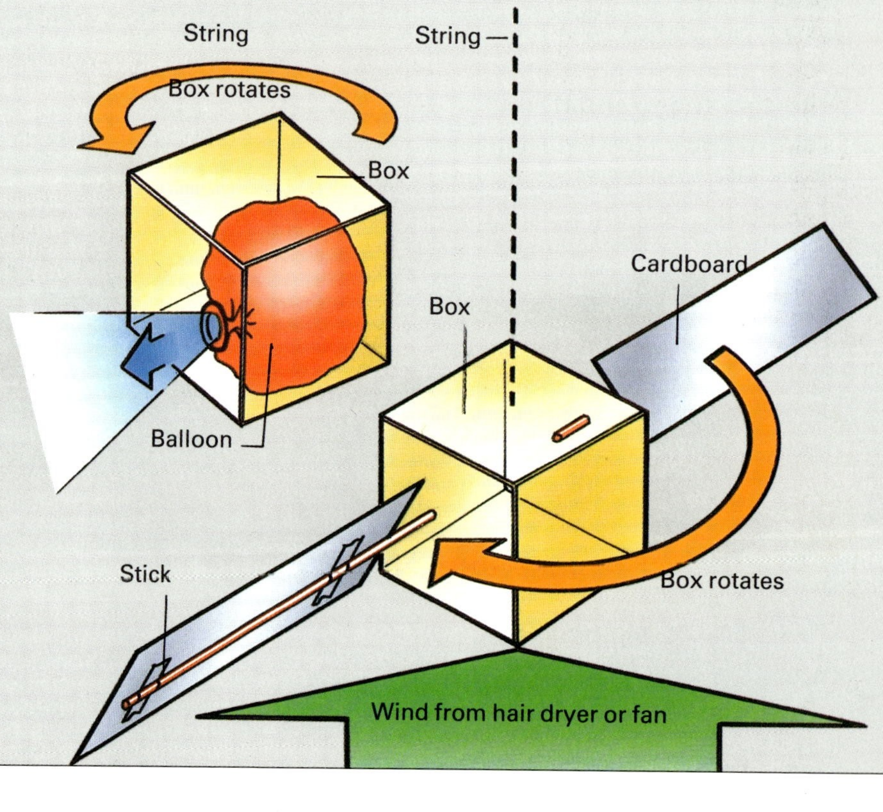

Transmitting

An antenna dish is a special shape that beams signals transmitted from the feedhorn. You can demonstrate this effect by covering a suitably shaped dish with silver foil, shiniest side up. Keep it as smooth as possible. A flash-light with a thin, powerful beam is best to act as the signal. By holding the light (representing the feedhorn) over the center of the dish and moving it around, you may be able to focus the beam into a circular area on a screen. Moving the light to one side and tilting the dish shows how an offset antenna works.

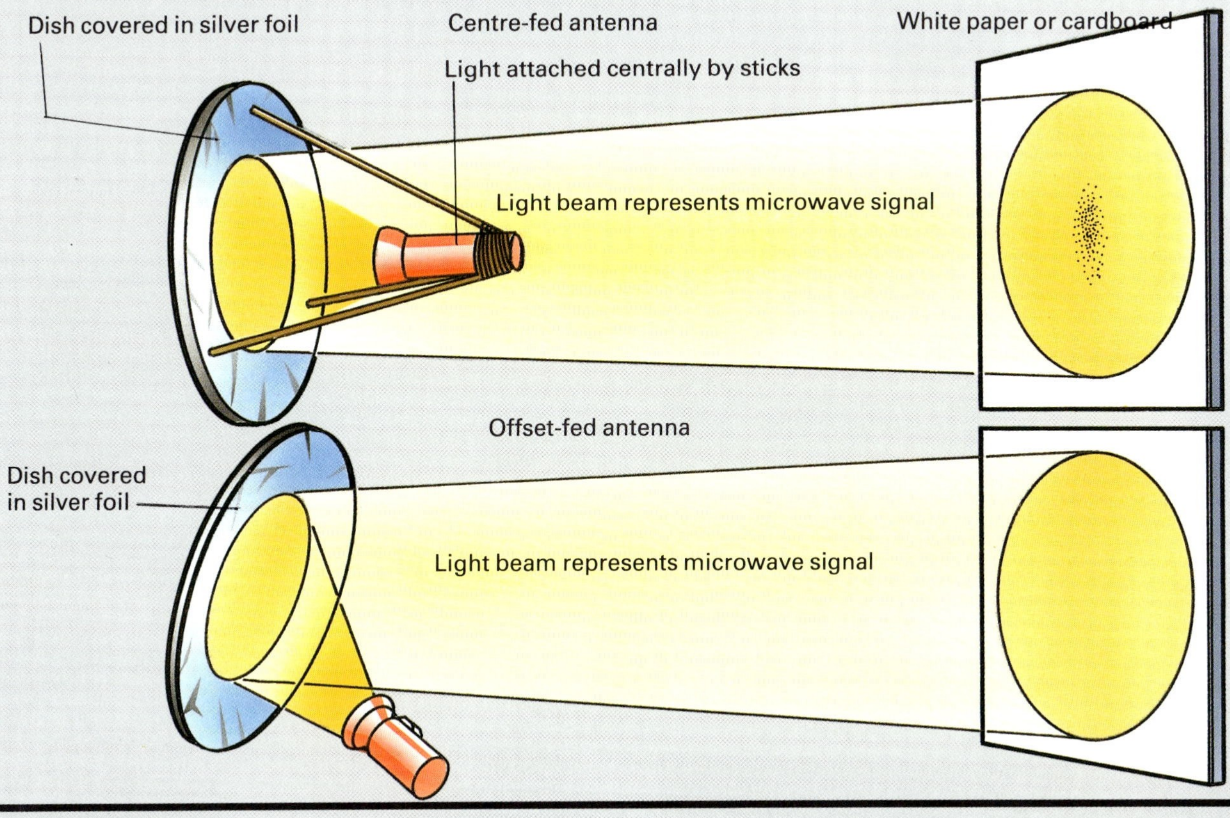

Footprints

A comsat's footprint is a complex combination of its dish shape and its electronics. However, the overall principle can be shown, again using a flashlight with a narrow beam. If it does not produce a narrow enough beam, a cardboard cone can be taped to the front. For the Earth, a model Earth globe is best, as beam shapes on real countries can be shown. But a baseball could be used instead. First, shine the light directly at the globe. This will produce a circular footprint. However, if the torch is shone at an angle to the globe, the circle becomes an oval, more like the shape of the simplest electronically-made footprint.

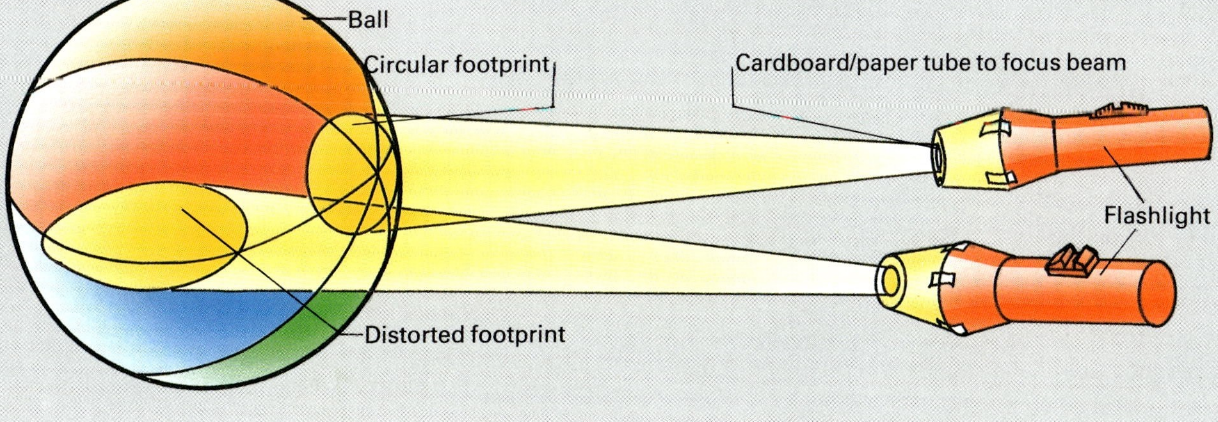

COMSAT ROUNDUP

Most of the comsats used now for day-to-day traffic come from American or European manufacturers. The U.S. still builds the most comsats, though Europe particularly Britain is catching up. Most satellites are built by several companies working together. One company is in overall charge – the prime contractor. Comsats like the Hughes *HS 376* or British Aerospace's *Olympus* are built to a standard design and then each one is tailored to an individual customer's requirements. The Soviet Union has its own system of comsats, including *Molniya*, the only comsat that is not in geostationary orbit. This is because most of the Soviet Union is at high latitudes that are difficult to reach from satellites over the Equator. To keep in touch with other countries, the Soviet Union also has a geosynchronous system based on its *Gorizont (Horizon)* satellites. In addition, China has developed its own communications satellites and can launch them with its own rockets.

A Soviet *Molniya* satellite

A *Gorizont (Horizon)* comsat

A Chinese comsat

Satellite (country)	Launch date
Score (U.S.A.)	12-18-58
Echo 1 (U.S.A.)	8-12-60
Courier (U.S.A.)	10-4-60
Telstar 1 (U.S.A.)	7-10-62
Relay 1 (U.S.A.)	12-13-62
Syncom 1 (U.S.A.)	2-14-63
Syncom 2 (U.S.A.)	7-16-63
Intelsat 1 (Early Bird) (U.S.A.)	4-6-65
Molniya 1 (U.S.S.R.)	4-23-65
DSCS 1 (U.S.A.)	6-16-66
ATS-1 (U.S.A.)	12-7-66
Tacsat (U.S.A.)	2-9-69
Skynet 1 (U.K.)	11-22-69
NATOsat 1 (U.S.A.)	3-20-70
FLTSATCOM (U.S.A.)	9-2-72
Westar 1 (U.S.A.)	4-13-74
ATS-6 (U.S.A.)	5-13-74
Oscar 7 (U.S.A.)	11-15-74
Molniya 3 (U.S.S.R.)	11-21-74
Symphonie (France/Germany)	12-19-74
Intelsat 4:F-1 (U.S.A.)	5-22-75
RCA Satcom 1 (U.S.A.)	12-13-75
Raduga (U.S.S.R.)	12-22-75
Marisat 1 (U.S.A.)	2-19-76
Comstar 1 (U.S.A.)	5-13-76
Palapa A1 (Indonesia)	7-8-76
Sakura 1/CS-1 (Japan)	12-15-77
OTS-2 (Europe)	5-11-78
Gorizont (U.S.S.R.)	7-5-79
SBS-1 (U.S.A.)	11-15-80
TDRS-1 (U.S.A.)	4-4-83
Anik C-2 (Canada)	6-18-83
Insat 1 (India)	8-30-83
China 15 (China)	4-8-84
Telecom 1 (France)	8-4-84
TDF-1 (France)	10-28-88
Skynet 4B (U.K.)	12-11-88
Olympus 1 (Europe)	due 1989

GLOSSARY

Apogee The point in an orbit around the Earth that is farthest from the center of the Earth.

Apogee motor A rocket engine fitted to a comsat to enable it to circularize its initial oval orbit.

Bus The main body of a satellite.

Clarke orbit Another name for a geostationary orbit, named after the British scientist and science-fiction writer, Arthur C. Clarke, who first proposed such an orbit in 1945.

DBS Direct Broadcast by Satellite. A system for transmitting television or radio programs to receiving dishes on individual homes.

Downlink The signal sent down from a comsat to receivers on the ground.

Elliptical orbit An "oval orbit" around the Earth. Only one type of comsat, the Soviet *Molniya*, uses this type of orbit.

ELV Expendable Launch Vehicle. A rocket such as *Ariane*, *Proton* or *Delta*.

Feedhorn A device at the focus of a dish antenna that collects or transmits signals.

Footprint The shape of the area on the Earth's surface where a signal from a satellite can be received.

Geostationary Earth Orbit (GEO) A circular orbit 36,000 km (22,000 mi) high over the Equator. Also called geosynchronous, synchronous, or Clarke Orbit.

Ground Station Equipment at a site on the Earth that transmits and receives signals from comsats.

Intelsat An international organization that operates comsats. Also the name of its series of comsats.

LEO Low Earth Orbit. A roughly circular orbit at a height of 160 to 1,000 km (100-600 mi) around the Earth.

PAM Payload Assist Module. A rocket stage designed by McDonnell Douglas for boosting comsats into orbit from a rocket or the Space Shuttle.

Perigee The point in an orbit around the Earth that is nearest to the Earth.

Solar cell A device that generates electricity directly from light. Hundreds of cells are arranged together on a spacecraft's solar panels to generate the power it requires during a mission. The rows of cells are called a solar array.

Spin stabilization Steadying a satellite by spinning its body so that it behaves like a gyroscope or spinning-top. Such satellites are almost always drum-shaped. Their antenna section is not spun, to keep it pointing at its ground stations.

Transfer orbit An elliptical orbit that transfers a satellite from LEO to GEO.

Transponder The equipment on board a comsat that receives, converts, and re-transmits signals. Comsats carry several transponders.

Three-axis stabilization Steadying a satellite by using gyroscopes and rocket thrusters.

Thrusters Tiny rocket engines used to make small changes to a satellite's position.

Uplink The signal sent up from ground stations to a comsat.

INDEX

Photographic Credits:
Cover and pages 4 (bottom left), 26 (both), 27 (both), 30 (all): Mat Irvine; pages 4 (top right) and 22: J. Allan Cash Photo Library; pages 15 and 21: ESA; pages 14 (inset) and 25 (both): NASA; page 17: British Aerospace; page 23: Vanessa Bailey.

[redacted]